STATISTICAL QUALITY CONTROL

STATISTICAL QUALITY CONTROL

M. Jeya Chandra

CRC Press
Taylor & Francis Group
Boca Raton London New York

CRC Press is an imprint of the
Taylor & Francis Group, an **informa** business

CRC Press
Taylor & Francis Group
6000 Broken Sound Parkway NW, Suite 300
Boca Raton, FL 33487-2742

First issued in paperback 2019

© 2001 by Taylor & Francis Group, LLC
CRC Press is an imprint of Taylor & Francis Group, an Informa business

No claim to original U.S. Government works

ISBN-13: 978-0-8493-2347-8 (hbk)
ISBN-13: 978-0-367-39725-8 (pbk)

**Visit the Taylor & Francis Web site at
http://www.taylorandfrancis.com**

**and the CRC Press Web site at
http://www.crcpress.com**

Preface

The objective of this book is to expose the reader to the various steps in the statistical quality control methodology. It is assumed that the reader has a basic understanding of probability and statistics taught at the junior level in colleges. The book is based on materials taught in a graduate-level course on statistical quality control in the Department of Industrial and Manufacturing Engineering at The Pennsylvania State University. The material discussed in this book can be taught in a 15-week semester and consists of nine chapters written in a logical manner. Some of the material covered in the book is adapted from journal publications. Sufficient examples are provided to illustrate the theoretical concepts covered.

I would like to thank those who have helped make this book possible. My colleague and friend, Professor Tom M. Cavalier of The Pennsylvania State University, has been encouraging me to write a textbook for the last ten years. His encouragement was a major factor in my writing this book. Many people are responsible for the successful completion of this book. I owe a lot to Professor Murray Smith of the University of Auckland, New Zealand, for his ungrudging help in generating the tables used in this book. My heartfelt thanks go to Hsu-Hua (Tim) Lee, who worked as my manager and helped me tremendously to prepare the manuscript; I would have been completely lost without his help. I would also like to thank Nicholas Smith for typing part of the manuscript and preparing the figures. Thanks are also due to Cecilia Devasagayam and Himanshu Gupta for their help in generating some of the end-of-chapter problems. I thank the numerous graduate students who took this course during the past few years, especially Daniel Finke, for their excellent suggestions for improvement.

A manuscript cannot be converted into a textbook without the help of a publisher. I would like to express my gratitude to CRC Press for agreeing to publish this book. My sincere thanks go to Cindy Renee Carelli, Engineering Acquisitions Editor at CRC Press, for her support in publishing this book. She was always willing to answer my questions and help me; publishers need persons like her to help authors. I also thank the anonymous reviewer of an earlier version of this manuscript for the excellent suggestions that led to substantial improvements of this manuscript.

Special gratitude and appreciation go to my wife, Emeline, and my children, Jean and Naveen, for the role they play in my life to make me a complete person. Finally, I thank my Lord and Savior, Jesus Christ, without whom I am nothing.

The Author

M. Jeya Chandra, Ph.D. is a professor of Industrial Engineering at The Pennsylvania State University, where he has been teaching for over 20 years. He has published over 50 papers in various journals and proceedings. In addition, he has won several teaching awards from the department, the College of Engineering, and the University. He has a B.E. in Mechanical Engineering from Madras University, India; an M.S. in Industrial Engineering from The Pennsylvania State University; and a Ph.D. in Industrial Engineering and Operations Research from Syracuse University.

Contents

I dedicate this book to

Mr. Sudarshan K. Maini, Chairman, Maini Group, Bangalore, India,

who was a great source of encouragement during the darkest period of my

professional life, and to his wonderful family.

1

Introduction

Quality can be defined in many ways, ranging from "satisfying customers' requirements" to "fitness for use" to "conformance to requirements." It is obvious that any definition of quality should include customers, satisfying whom must be the primary goal of any business. Experience during the last two decades in the U.S. and world markets has clearly demonstrated that quality is one of the most important factors for business success and growth. Businesses achieving higher quality in their products enjoy significant advantage over their competition; hence, it is important that the personnel responsible for the design, development, and manufacture of products understand properly the concepts and techniques used to improve the quality of products. Statistical quality control provides the statistical techniques necessary to assure and improve the quality of products.

Most of the statistical quality control techniques used now have been developed during the last century. One of the most commonly used statistical tools, *control charts,* was introduced by Dr. Walter Shewart in 1924 at Bell Laboratories. The *acceptance sampling* techniques were developed by Dr. H. F. Dodge and H. G. Romig in 1928, also at Bell Laboratories. The use of *design of experiments* developed by Dr. R. A. Fisher in the U.K. began in the 1930s. The end of World War II saw increased interest in quality, primarily among the industries in Japan, which were helped by Dr. W. E. Deming. Since the early 1980s, U.S. industries have strived to improve the quality of their products. They have been assisted in this endeavor by Dr. Genichi Taguchi, Philip Crosby, Dr. Deming, and Dr. Joseph M. Juran. Industry in the 1980s also benefited from the contributions of Dr. Taguchi to *design of experiments, loss function,* and *robust design.* The recent emphasis on teamwork in design has produced *concurrent engineering.* The standards for a quality system, ISO 9000, were introduced in the early 1990s. They were later modified and enhanced substantially by the U.S. automobile industries, resulting in QS-9000.

The basic steps in statistical quality control methodology are represented in Figure 1.1, which also lists the output of each step. This textbook covers most of the steps shown in the figure. It should be emphasized here that the steps given are by no means exhaustive. Also, most of the activities must be performed in a parallel, not sequential, manner. In Chapter 2, Tolerancing, assembly tolerance is allocated to the components of the assembly. Once tolerances

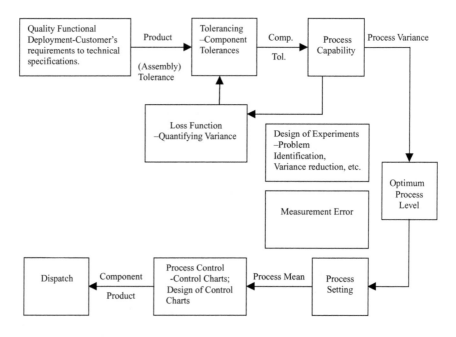

FIGURE 1.1
Quality control methodology.

on the quality characteristics of the components are determined, processes must be selected for manufacture of the components. The personnel responsible for process selection must be cognizant of the effect of quality characteristic variances on the quality of the product. This process, developed by Dr. Taguchi, is discussed in Chaper 3, Loss Function. Robust design, which is based upon loss function, is also discussed in this chapter. Process capability analysis, which is an important step for selection of processes for manufacture of the components and the product, is discussed in Chapter 4. Process capability analysis cannot be completed without ascertaining that the process is in control. Even though this is usually achieved using control charts, this topic is covered later in the book. The effect of measurement error, which is addressed in Chapter 5, should also be taken into consideration. Emphasis in the text is given to modeling of errors, estimation of error variances, and the effect of measurement errors on decisions related to quality. After process selection is completed, optimal means for obtaining the quality characteristics must be determined, and these are discussed in Chapter 6, Optimal Process Levels. The emphasis in this chapter is on the methodologies used and the development of objective functions and solution procedures used by various researchers. The next step in the methodology is process setting, as discussed in Chapter 7, in which the actual process mean is brought as close as possible to the optimal value determined earlier. Once the process setting is completed, manufacture of the components can begin. During the entire period of manufacture, the mean and variance of the process must be kept at their respective target values, which is accomplished, as described in Chapter 8, through process control, using control charts. Design of experiments, discussed in Chapter 9, can be used in any of the steps mentioned earlier. It serves as a valuable tool for identifying causes of problem areas, reducing variance, determining the levels of process parameters to achieve the target mean, and more.

Many of the steps described must be combined into one larger step. For example, concurrent engineering might combine tolerancing, process selection, robust design, and optimum process level into one step. It is emphasized again that neither the quality methodology chart in Figure 1.1 nor the treatment of topics in this book implies a sequential carrying out of the steps.

2

Tolerancing

CONTENTS

2.1   Introduction

In mass production, products are assembled using parts or components manufactured or processed on different processes or machines. This requires complete interchangeability of parts while assembling them. On the other hand, there will always be variations in the quality characteristics (length, diameter, thickness, tensile strength, etc.) because of the inherent variability introduced by the machines, tools, raw materials, and human operators. The presence of unavoidable variation and the necessity of interchangeability require that some limits be specified for the variation of any quality characteristic. These allowable variations are specified as *tolerances*. Usually, the tolerances on the quality characteristics of the final assembly/product are specified by either the customer directly or the designer based upon the functional requirements specified by the customer. The important next step is to allocate these assembly tolerances among the quality characteristics of the components of the assembly. In this chapter, we will learn some methods that have been developed for tolerance allocation among the components.

2.2   Preliminaries

We will consider assemblies consisting of k components ($k \geq 2$). The quality characteristic of component i that is of interest to the designer (user) is denoted by X_i. This characteristic is assumed to be of the Nominal-the-Better type. The upper and lower specification limits of X_i are U_i (USL$_i$) and L_i (LSL$_i$), respectively.

The assembly quality characteristic of interest to the designer (user) denoted by X is a function of X_i, $i = 1, 2, \ldots, k$. That is,

$$X = f(X_1, X_2, \ldots, X_k) \tag{2.1}$$

At first, we will consider linear functions of X_i only:

$$X = X_1 \pm X_2 \pm X_3 \pm \cdots \pm X_k \tag{2.2}$$

The upper and lower specification limits of X are U (USL) and L (LSL), respectively. These are assumed to be given by the customer or determined by the designer based on the functional requirements specified by the customer. Some examples of the assemblies with linear relationships among the assembly characteristics and component characteristics are given next.

Example 2.1

Three different assemblies are given in Figures 2.1a, b, and c. In the shaft and sleeve assembly shown in Figure 2.1a, the inside diameter of the sleeve and

a.

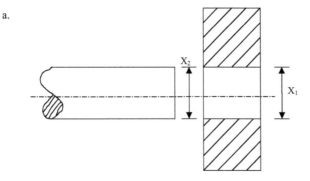

b.

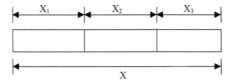

c.

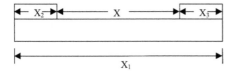

FIGURE 2.1
Some examples of assemblies.

the outside diameter of the shaft are the *component characteristics*, and the clearance between these diameters is the *assembly characteristic*. Let X_1 and X_2 represent the inside diameter of the sleeve and the outside diameter of the shaft, respectively, and let X denote the clearance between these two diameters. Then, the relationship between the assembly characteristic and the component characteristic is given by:

$$X = X_1 - X_2 \tag{2.3}$$

In the assembly given in Figure 2.1b, the component characteristics are the lengths of these components, denoted by X_1, X_2, and X_3, and X is the length of the assembly. The relationship among the tolerances in this case is given by:

$$X = X_1 + X_2 + X_3 \tag{2.4}$$

In Figure 2.1c, the assembly characteristic X is related to the component characteristics X_1, X_2, and X_3 as:

$$X = X_1 - X_2 - X_3 \qquad (2.5)$$

In general, these relations can be written as in Eq. (2.2).

2.3   Additive Relationship

Tolerance is the difference between the upper and lower specification limits. Let the tolerance of X_i be T_i, $i = 1, 2, \ldots, k$, and let the tolerance of the assembly characteristic X be T. Then,

$$T_i = U_i - L_i, \quad i = 1, 2, \ldots, k \qquad (2.6)$$

where L_i and U_i are the lower and upper specification limits of characteristic X_i, respectively. Similarly,

$$T = U - L, \qquad (2.7)$$

where L and U are the lower and upper specification limits of X, respectively.
The relationship between T and $T_1, \ldots, T_k$ can now be derived using the assembly in Figure 2.1c as an example. The relationship among the tolerances was given in Eq. (2.5) as:

$$X = X_1 - X_2 - X_3$$

As U is the maximum allowable value of X, it is realized when X_1 is at its maximum allowable value and X_2 and X_3 are at their respective minimum allowable values. Hence,

$$U = U_1 - L_2 - L_3 \qquad (2.8)$$

Similarly L, being the minimum allowable value of X, is obtained when X_1 is at its minimum allowable value and X_2 and X_3 are, respectively, at their maximum allowable values. Hence,

$$L = L_1 - U_2 - U_3 \qquad (2.9)$$

Now, as per Eq. (2.7),

$$T = (U - L)$$

which can be written using Eqs.(2.8) and (2.9) as:

$$T = (U_1 - L_2 - L_3) - (L_1 - U_2 - U_3)$$
$$= (U_1 - L_1) + (U_2 - L_2) + (U_3 - L_3)$$
$$= T_1 + T_2 + T_3 \tag{2.10}$$

In general, for any linear function $X = X_1 \pm X_2 \pm X_3 \pm \cdots \pm X_k$,

$$T = T_1 + T_2 + T_3 + \cdots + T_k \tag{2.11}$$

This is called an *additive relationship*. The design engineer can allocate tolerances $T_1, \ldots, T_k$ among the k components, for a given (specified) T, using this additive relationship. Let us now use this relationship in an example to allocate tolerance among the components.

Example 2.2

Let us consider the assembly depicted in Figure 2.1a, having two components (sleeve and shaft) with characteristics (diameters) X_1 and X_2, respectively. The assembly characteristic is the clearance between the sleeve and the shaft, denoted by X, which is equal to:

$$X = X_1 - X_2 \tag{2.12}$$

and

$$T = T_1 + T_2 \tag{2.13}$$

Let us assume that the tolerance on X, which is T, is 0.001 in. Using Eq. (2.13), we get:

$$T_1 + T_2 = 0.001 \tag{2.14}$$

There are two unknowns, T_1 and T_2, and only one equation. In general, if the assembly has k components, there will be k unknowns and still only one equation. We need $(k-1)$ more equations or relations among the components' tolerances, T_i's, in order to solve for them. These relations usually reflect the difficulties associated with maintaining these tolerances while machining/processing the components. As we will see later, the manufacturing cost decreases when the tolerance on the quality characteristic increases. Let us assume that, in our example, the difficulty levels of maintaining both T_1 and T_2 are the same, hence the designer would like these tolerances to be equal. That is,

$$T_1 = T_2 \tag{2.15}$$

Using (2.14) and (2.15), we obtain

$$T_1 = T_2 = \frac{T}{2} = \frac{0.001}{2} = 0.0005$$

On the other hand, if it is more difficult to process component 1 than component 2, then the designer would like to have T_1 greater than T_2. For example, the following relation can be used:

$$T_1 = 2T_2 \tag{2.16}$$

In this case, using Eqs. (2.14) and (2.16), we get:

$$2T_2 + T_2 = 0.001 \rightarrow T_2 = 0.00033$$
$$T_1 = 0.00066$$

rounding off to five decimal places. It may be noted here that the number of decimal places carried in the tolerance values depends upon the precision of the instruments/gauges used to measure the characteristics.

2.4   Probabilistic Relationship

As this relationship depends upon the probabilistic properties of the component and assembly characteristics, it necessary to make certain *assumptions* regarding these characteristics:

1. X_i's are independent of each other.
2. Components are randomly assembled.
3. $X_i \sim N(\mu_i, \sigma_i^2)$; that is, the characteristic X_i is normally distributed with a mean μ_i and a variance σ_i^2 (this assumption will be relaxed later on).
4. The process that generates characteristic X_i is adjusted and controlled so that the mean of the distribution of X_i, μ_i, is equal to the nominal size of X_i, denoted by B_i, which is the mid-point of the tolerance region of X_i. That is,

$$\mu_i = \frac{(U_i + L_i)}{2} \tag{2.17}$$

5. The standard deviation of the distribution of the characteristic X_i, generated by the process, is such that 99.73% of the characteristic X_i

TABLE 2.1

Areas for Different Ranges Under Standard Normal Curve

Range	% Covered within the Range	% Outside the Range	Parts per million Outside the Range
$(\mu - 1\sigma)$ to $(\mu + 1\sigma)$	68.26	31.74	317,400
$(\mu - 2\sigma)$ to $(\mu + 2\sigma)$	95.44	4.56	45,600
$(\mu - 3\sigma)$ to $(\mu + 3\sigma)$	99.73	0.27	2700
$(\mu - 4\sigma)$ to $(\mu + 4\sigma)$	99.99366	0.00634	63.4
$(\mu - 5\sigma)$ to $(\mu + 5\sigma)$	99.9999426	0.0000574	0.574
$(\mu - 6\sigma)$ to $(\mu + 6\sigma)$	99.9999998	0.0000002	0.002

falls within the specification limits for X_i. Based upon the property of normal distribution, this is represented as (see Table 2.1):

$$U_i - L_i = T_i = 6\sigma_i, \quad i = 1, 2, \ldots, k \tag{2.18}$$

Let μ and σ^2 be the mean and variance, respectively, of X. As $X = X_1 \pm X_2 \pm X_3 \pm \cdots \pm X_k$,

$$\mu = \mu_1 \pm \mu_2 \pm \mu_3 \pm \cdots \pm \mu_k \tag{2.19}$$

and as the X_i's are independent of each other,

$$\sigma^2 = \sigma_1^2 + \sigma_2^2 + \cdots + \sigma_k^2 \tag{2.20}$$

Because of assumption 2 (above), the assembly characteristic X is also normally distributed.

Let us assume that 99.73% of all assemblies have characteristic X within the specification limits U and L. This yields a relation similar to Eq. (2.18):

$$(U - L) = T = 6\sigma \tag{2.21}$$

From Eqs. (2.18) and (2.21), we get:

$$\sigma_i^2 = \left(\frac{T_i}{6}\right)^2, \quad i = 1, 2, \ldots, k \tag{2.22}$$

and

$$\sigma^2 = \left(\frac{T}{6}\right)^2 \tag{2.23}$$

Combining Eqs. (2.20), (2.22), and (2.23) yields:

$$\left(\frac{T}{6}\right)^2 = \left(\frac{T_1}{6}\right)^2 + \left(\frac{T_2}{6}\right)^2 + \cdots + \left(\frac{T_k}{6}\right)^2 \tag{2.24}$$

or

$$T = \sqrt{T_1^2 + T_2^2 + \cdots + T_k^2} \tag{2.25}$$

The relationship given in Eq. (2.25) is called a *probabilistic relationship* and provides another means for allocating tolerances among components for a given assembly tolerance, T. Let us use this relationship to allocate tolerances among the two components of the assembly considered earlier.

Example 2.3

We may recall that by using the additive relationship (and assuming that $T_1 = T_2$), the tolerances were obtained as $T_1 = T_2 = 0.0005$ in. Now setting $T = 0.001$ in Eq. (2.25) yields:

$$\sqrt{T_1^2 + T_2^2} = 0.0005$$

We face the same problem we encountered earlier; that is, we have only one equation, whereas the number of variables is 2 (in general, it is k). If we introduce the same first relation used earlier ($T_1 = T_2$), then Eq. (2.26) gives:

$$\sqrt{2T_1^2} = 0.001 \rightarrow T_1 = \frac{0.001}{\sqrt{2}}$$

$$T_1 = T_2 = 0.00071$$

if five significant digits are kept after the decimal point.

The component tolerances T_1 and T_2 obtained using the additive and probabilistic relationships for the same assembly tolerance $T = 0.001$ and the same relationship between T_1 and T_2 ($T_1 = T_2$) are summarized in Table 2.2.

It can be seen that the probabilistic relationship yields larger values for the component tolerances compared to the additive relationship (43% more in this example). We will examine the advantages and disadvantages of this increase in component tolerances next.

Now, we have two relations between T and $(T_1, \ldots, T_k)$:

$$T = T_1 + T_2 + T_3 + \cdots + T_k \tag{2.26}$$

TABLE 2.2

Comparison of Additive and Probabilistic Relationships

	Additive	Probabilistic
T_1	0.0005	0.00071
T_2	0.0005	0.00071

and

$$T = \sqrt{T_1^2 + T_2^2 + \cdots + T_k^2} \qquad (2.27)$$

Let us denote T in (2.26) by T_a and T in (2.27) by T_p (T_{a_i} and T_{p_i} for components); then:

$$T_a = T_{a_1} + T_{a_2} + \cdots + T_{a_k} \qquad (2.28)$$

and

$$T_p = \sqrt{T_{p_1}^2 + T_{p_2}^2 + \cdots + T_{p_k}^2} \qquad (2.29)$$

In Examples (2.2) and (2.3), we set $T_a = T_p = 0.001$ and solved for $T_{a_1} = T_{a_2} = 0.005$ and $T_{p_1} = T_{p_2} = 0.00071$. We saw that $T_{p_1} > T_{a_1}$ and $T_{p_2} > T_{a_2}$. Now let us examine the advantages and disadvantages of using the probabilistic relationship to allocate tolerances among the components.

2.4.1 Advantages of Using Probabilistic Relationship

It is a well-established fact that manufacturing cost decreases as the tolerance on the quality characteristic increases, as shown in Figure 2.2. Hence, the manufacturing cost of the components will decrease as a result of using the probabilistic relationship.

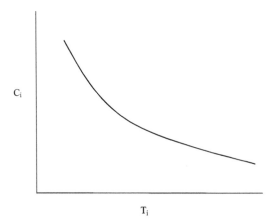

FIGURE 2.2
Curve showing cost–tolerance relationship.

2.4.2   Disadvantages of Using Probabilistic Relationship

If the probabilistic relationship is used, then the tolerance on the internal diameter of the sleeve and the outside diameter of the shaft is 0.00071. This implies that the maximum allowable range of the internal diameter of the sleeve is 0.00071. Likewise, the maximum allowable range of the outside diameter of the shaft is also 0.00071. Hence, the actual maximum range of the clearance of the assemblies assembled using these components will be

$$T_1 + T_2 = 0.00071 + 0.00071 = 0.000142$$

The allowable range of the clearance of the assemblies, T, is 0.001. This will obviously lead to rejection of the assemblies. In order to estimate the actual proportion of rejection, we need the probability distribution of the assembly characteristic, X, along with its mean and standard deviation.

If the component characteristics are normally distributed, then the assembly characteristic is also normally distributed. If the means of the component characteristics are equal to their respective nominal sizes, then the mean of the assembly characteristic is equal to the assembly nominal size. The only equation that contains the variance, σ^2, is

$$\sigma^2 = \sigma_1^2 + \sigma_2^2 + \cdots + \sigma_k^2 \tag{2.30}$$

The standard deviations, σ_1, and σ_2, are (per assumption (5) made earlier):

$$\sigma_1 = \frac{T_1}{6} = \frac{0.00071}{6} = 0.000118$$

and

$$\sigma_2 = \frac{T_2}{6} = 0.000118$$

Hence,

$$\sigma = \sqrt{0.000118^2 + 0.00118^2} = 0.000167$$

and

$$6\sigma = 6 \times 0.000167$$

$$= 0.001$$

Because X is normally distributed, the range 6σ contains 99.73% of the values of X, which is the assembly characteristic (see Table 2.1). Hence, the percentage rejection of the assemblies is <0.27%. This is illustrated in Figure 2.3.

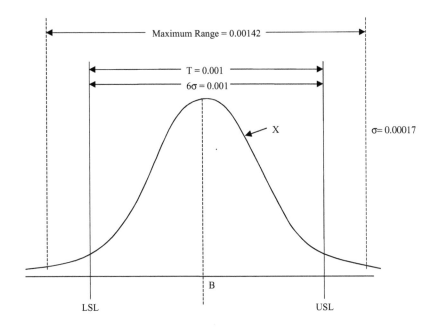

FIGURE 2.3
The result of a probabilistic relationship.

Now, let us compare the percentage of rejection of the assemblies when the component tolerances are determined using the additive relationship. Now the standard deviations, σ_1 and σ_2 are

$$\sigma_1 = \frac{0.0005}{6} = 0.0000833$$

and

$$\sigma_2 = 0.0000833$$

Hence,

$$\sigma = \sqrt{2 \times 0.0000833^2} = 0.0001179$$

and

$$6\sigma = 6 \times 0.000117 = 0.00071$$

As 6σ is less than the maximum allowable range ($T = 0.001$), the percentage rejection now is $\cong 0$. This is illustrated in Figure 2.4.

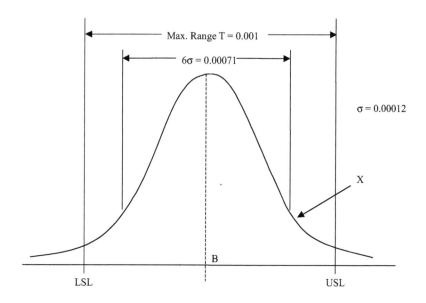

FIGURE 2.4
The result of an additive relationship.

   Thus, determining component tolerances using the probabilistic relationship increases the percentage rejection of assemblies while decreasing the
manufacturing cost of the components. It also increases inspection cost (100%
inspection of assemblies).

2.4.3   Probabilistic Relationship for Non-Normal Component Characteristics

Let the probability density function of X_i be $f_i(x_i)$ with a mean μ_i and a variance σ_i^2. We assume that the range that contains 100% or close to 100% of all
possible values of X_i is $g_i\sigma_i$. It is still assumed that:

$$T_i = g_i\sigma_i \tag{2.31}$$

(ideally $T_i >>>> g_i\sigma_i$). This can be written as:

$$\sigma_i = \frac{T_i}{g_i} \tag{2.32}$$

Now, given that $X = X_1 \pm X_2 \pm X_3 \pm \cdots \pm X_k$, the distribution of X is approximately normal, because of the Central Limit Theorem. So,

$$T_p = 6\sigma \rightarrow \sigma = \frac{T_p}{6},$$

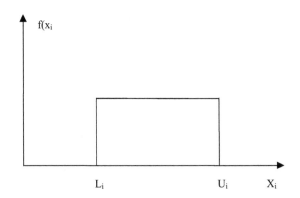

FIGURE 2.5
Uniform distribution.

assuming 99.73% coverage. Using the formula $\sigma^2 = \sigma_1^2 + \sigma_2^2 + \cdots + \sigma_k^2$,

$$\left(\frac{T_p}{6}\right)^2 = \left(\frac{T_1}{g_1}\right)^2 + \left(\frac{T_2}{g_2}\right)^2 + \cdots + \left(\frac{T_k}{g_k}\right)^2$$

$$T_p = 6 \times \sqrt{\left(\frac{T_1}{g_1}\right)^2 + \left(\frac{T_2}{g_2}\right)^2 + \cdots + \left(\frac{T_k}{g_k}\right)^2} \qquad (2.33)$$

2.4.3.1   Uniform Distribution
If $f_i(x_i)$ is a uniform distribution for all i, then (Figure 2.5):

$$T_i = U_i - L_i \qquad (2.34)$$

$$\mathrm{Var}(X_i) = \frac{(\mathrm{Range})^2}{12}$$

$$= \frac{(U_i - L_i)^2}{12}$$

$$\sigma_i^2 = \frac{(T_i)^2}{12} \qquad (2.35)$$

$$T_i = \sqrt{12}\, \sigma_i \quad (g_i = \sqrt{12}) \qquad (2.36)$$

$$T_p = 6\sqrt{\frac{T_1^2}{12} + \frac{T_2^2}{12} + \cdots + \frac{T_k^2}{12}}$$

$$= \sqrt{3}\sqrt{T_1^2 + T_2^2 + \cdots + T_k^2} \qquad (2.37)$$

The assembly characteristic, X, can be assumed to be approximately normally distributed, and the probability tolerance relationship in Eq. (2.30) can be used, even if the component characteristics are uniformly distributed, when the number of components, k, is large, because of the Central Limit Theorem. However, this approximation will yield poor results when k is small, as illustrated by the following example.

Example 2.4

Consider an assembly consisting of two components with quality characteristics X_1 and X_2. The assembly characteristic X is related to X_1 and X_2 as follows:

$$X = X_1 + X_2$$

Assume that it is possible to select processes for manufacturing the components such that X_1 and X_2 are uniformly distributed. The ranges of X_1 and X_2 are (L_1, U_1) and (L_2, U_2), respectively. The tolerance on the assembly characteristic is specified as 0.001. Allocate this tolerance among the components using the probabilistic relationship, assuming that the component tolerances are equal.

If we use the additive relationship, the tolerances T_1 and T_2 are

$$0.001 = T_1 + T_2$$

$$T_1 = T_2 = \frac{0.001}{2} = 0.0005$$

Now let us use the probabilistic relationship:

$$T_p = \sqrt{3}\sqrt{T_1^2 + T_2^2}$$

$$0.001 = \sqrt{3}\sqrt{2T_1^2} \rightarrow T_1(\sqrt{6}) = 0.001$$

$$T_1 = T_2 = 0.000408 \cong 0.00041.$$

It can be seen that the probabilistic relationship yields tolerances that are smaller than the tolerances obtained using the additive relationship. The reason for this is that the assembly characteristic X is not normally distributed. The Central Limit Theorem is true only for large values of k.

The actual distribution of X is a triangular distribution. Figure 2.6 contains this distribution, obtained as result of adding two independent uniform random variables with the same minimum and maximum limits (that is, assuming $L_1 = L_2$ and $U_1 = U_2$). Here, the range containing 100% of X is

$$2U_1 - 2L_1 = 2(U_1 - L_1) = 2T_1$$

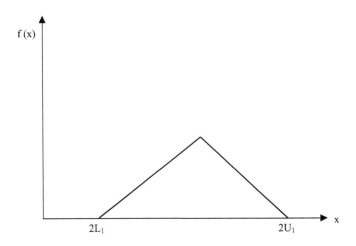

FIGURE 2.6
Sum of two uniform distributions.

Hence, the correct probabilistic relationship is

$$T_p = 2T_1, \quad \text{not } T_p = \sqrt{6}T_1$$

This example highlights the problem in using the probabilistic relationship when the component characteristics are not normally distributed and when k is small.

Example 2.5

Now let us assume that the number of components is 10 (that is, $k = 10$ instead of 2). We assume that the component characteristics are uniformly distributed in the ranges of the respective specification intervals and that the tolerances (T_i) are all equal.

The additive relationship yields:

$$10T_1 = 0.001 \rightarrow T_1 = T_2 = \cdots T_{10} = 0.0001$$

The probabilistic relationship yields

$$\sqrt{3}\sqrt{10T_1^2} = 0.001 \rightarrow T_1\sqrt{30} = 0.001$$

$$T_1 = T_2 = \cdots T_{10} = \frac{0.001}{\sqrt{30}} = 0.000183$$

In this case, the tolerances obtained using the probabilistic relationship are larger than the tolerances resulting from the additive relationship. This is because the distribution of X is closer to the normal distribution as k is large (10).

2.4.3.2 *Beta Distribution*

The beta distribution is a more flexible probability density function compared to the normal distribution because it can accommodate different
ranges (not always from $-\infty$ to $+\infty$, as in the normal distribution) and different shapes from left skewed to symmetrical to right skewed (not always
symmetric, as in the normal distribution). It has four parameters: a and b,
which are, respectively, the minimum and maximum values that a beta random variable can assume, and γ and η, which are the shape parameters. The
density function is

$$f(x) = \frac{1}{(b-a)B(\gamma,\eta)}\left[\frac{x-a}{b-a}\right]^{(\gamma-1)}\left[1-\frac{x-a}{b-a}\right]^{(\eta-1)}, \quad a \le x \le b \qquad (2.38)$$

where

$$B(\gamma,\eta) = \frac{\Gamma(\gamma)\Gamma(\eta)}{\Gamma(\gamma+\eta)} = \int_0^1 v^{\gamma-1}(1-v)^{\eta-1}dv \qquad (2.39)$$

and

$$\Gamma(\gamma) = (\gamma-1)! \qquad (2.40)$$

Though the density function is not a simple function, the flexibility it offers
and its finite range make it an excellent candidate for representing many
quality characteristics in real life. The shape of the density function depends
upon the values of the shape parameters γ and η.

The mean and variance of the beta distribution are given next. The shape
parameters γ and η are also expressed as functions of the mean and variance below. Finally, the range of the beta distribution is expressed in terms
of the standard deviation.

$$E(X) = \mu = a + (b-a)\frac{\gamma}{(\gamma+\eta)} = \frac{(b\gamma+a\eta)}{(\gamma+\eta)} \qquad (2.41)$$

$$\mathrm{Var}(X) = \sigma^2 = \frac{(b-a)^2\gamma\eta}{(\gamma+\eta+1)(\gamma+\eta)^2} \qquad (2.42)$$

$$\gamma = \left[\frac{(\mu-a)^2(b-\mu)-\sigma^2(\mu-a)}{\sigma^2(b-a)}\right] \qquad (2.43)$$

$$\eta = \left[\frac{(\mu - a)(b - \mu) - \sigma^2(b - \mu)}{\sigma^2(b - a)}\right] \tag{2.44}$$

$$(b - a) = \text{Range} = t_i = \sqrt{\frac{(\gamma + \eta + 1)(\gamma + \eta)^2}{\gamma \eta}} \sigma_i \tag{2.45}$$

It can be seen from Eq. (2.42) that for the beta distribution:

$$g_i = \sqrt{\frac{(\gamma + \eta + 1)(\gamma + \eta)^2}{\gamma \eta}} \tag{2.46}$$

Hence, for component characteristics that follow the beta distribution, the probabilistic relationship in Eq. (2.30) becomes:

$$T_p = 6\sqrt{\frac{\gamma \eta}{(\gamma + \eta + 1)(\gamma + \eta)^2}} \sqrt{T_1^2 + T_2^2 + \cdots + T_k^2} \tag{2.47}$$

The formulas we derived so far are based on the following *assumptions:*

1. The probability distributions of the quality characteristics generated by the processes are known.
2. The *capability* of the process matches the (engineering) specification tolerance (T_i) of the quality characteristic X_i. (Here, capability means the range of all possible values of the quality characteristic generated by the process.) In other words, the range of all possible values of quality characteristic X_i is equal to the range of allowable values as per specifications (that is, the range $U_i - L_i$):

$$(T_i = g_i \sigma_i)$$

3. The mean of the distribution of the characteristic X_i generated by the process μ_i is equal to the nominal size, B_i, which is the midpoint of the tolerance interval. That is,

$$\mu_i = B_i = \frac{(U_i + L_i)}{2}, \quad \text{for } i = 1, 2, \ldots, k$$

Let us consider assumption (2). The range of all *possible* values of any quality characteristic generated by a process is called the *natural process tolerance* of that process and is denoted by t_i. This is different from T_i, which is the *allowable* range of X_i:

$$t_i = g_i \sigma_i \tag{2.48}$$

TABLE 2.3

Natural Process Tolerances

Distributions	$t_i(g_i\sigma_i)$	
Normal	$6\sigma_i$	(99.73%)
Uniform	$\sqrt{12}\,\sigma_i$	(100%)
Beta	$\sqrt{\dfrac{(\gamma+\eta+1)(\gamma+\eta)^2}{\gamma\eta}}\,\sigma_i$	(100%)

The range of *allowable* values of the quality characteristic X_i is the *engineering tolerance*, which is denoted by T_i. Per assumption (2), $T_i = t_i$ (ideally, $T_i \geq t_i$). Table 2.3 gives the natural process tolerances for the normal, uniform, and beta distributions obtained earlier.

The natural process tolerance is called *process capability* in real-life applications. For a given machine or process, t_i depends upon the operation performed on the machine (such as turning, forming, etc.) and the material used (brass, stainless steel, etc.).

Assumption (3) is not realistic, because in real-life applications it is very difficult to make the mean of X_i, μ_i, exactly equal to the nominal size, B_i. It requires a long time to achieve this, hence industries allow a maximum nonzero deviation between the mean and the nominal size. In the next section, we will consider the component tolerance allocation problem in cases where the process means (μ_i) are not equal to the respective nominal sizes (B_i), which are assumed to be the mid-points of the respective tolerance intervals. That is,

$$\mu_i \neq B_i\left(\mu_i \neq \frac{L_i + U_i}{2}\right)$$

2.5 Tolerance Allocation when the Means Are Not Equal to the Nominal Sizes

The method of component tolerance allocation for a given assembly tolerance discussed in this section is based on the paper, "The Application of Probability to Tolerances Used in Engineering Designs."[6] The following *assumptions* are made:

1. X_i's are independent of each other.
2. Components are randomly assembled.
3. $X_i \sim N(\mu_i, \sigma_i^2)$; that is, the characteristic X_i is normally distributed with a mean μ_i and a variance σ_i^2.

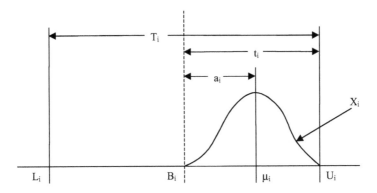

FIGURE 2.7
Right shift in the mean of X_i.

4. The standard deviation of the distribution of the characteristic X_i generated by the process is such that the natural process tolerance of $X_i(= t_i)$ is less than $T_i(= U_i - L_i)$. That is, the *actual* range of all possible values of X_i is contained within the *allowable* range of the values of X_i.

5. The relationship between the assembly characteristic and the component characteristic is assumed to be

$$X = X_1 \pm X_2 \pm X_3 \pm \cdots \pm X_k$$

The location of the mean of the distribution of X_i, μ_i, is fixed while setting the process. Let the *maximum allowable* difference between the mean μ_i and the midpoint B_i (nominal size of X_i) of the tolerance region be a_i.

Case 1, when $\mu_i > B_i$. Let us consider the worst case, when μ_i is at its maximum allowable location, for $i = 1, 2, ..., k$. From Figure 2.7:

$$U_i = B_i + \frac{T_i}{2} \rightarrow B_i = U_i - \frac{T_i}{2}$$

$$= \mu_i + \frac{t_i}{2} \rightarrow \mu_i = U_i - \frac{t_i}{2}$$

$$\mu_i - B_i = \frac{T_i}{2} - \frac{t_i}{2} \tag{2.49}$$

$$a_i = \frac{(T_i - t_i)}{2} \tag{2.50}$$

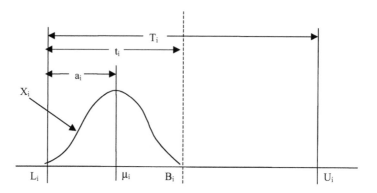

FIGURE 2.8
Left shift in the mean of X_i.

Case 2, when $\mu_i < B_i$. Again, let us consider the worst case, when μ_i is at its minimum allowable location, for $i = 1, 2, \ldots, k$. From Figure 2.8:

$$L_i = B_i - \frac{T_i}{2} \rightarrow B_i = L_i + \frac{T_i}{2}$$

$$= \mu_i - \frac{t_i}{2} \rightarrow \mu_i = L_i + \frac{t_i}{2}$$

$$B_i - \mu_i = \frac{T_i}{2} - \frac{t_i}{2} \tag{2.51}$$

$$a_i = \frac{(T_i - t_i)}{2} \tag{2.52}$$

Now let us consider the assembly characteristic X. As we do not know the exact location of μ_i anymore (all we know is that maximum $|\mu_i - B_i| = a_i$), we do not know the exact location of μ either. Let the maximum and minimum possible values of μ be μ_{max} and μ_{min}, respectively (Figure 2.9). In the figure:

$$\mu_{max} - \mu_{min} = 2\sum_{i=1}^{k} a_i \tag{2.53}$$

At any given position of μ in the range (μ_{min}, μ_{max}), the range containing all possible (99.73% for normal distribution) values of X is

$$t = 6\sigma = 6\sqrt{\sigma^2} = 6\sqrt{\sigma_1^2 + \sigma_2^2 + \cdots + \sigma_k^2}$$

$$= 6\sqrt{\left(\frac{t_1}{6}\right)^2 + \left(\frac{t_2}{6}\right)^2 + \cdots + \left(\frac{t_k}{6}\right)^2}$$

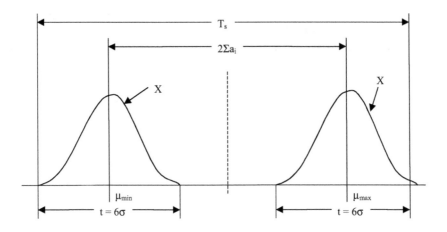

FIGURE 2.9
Distribution of X.

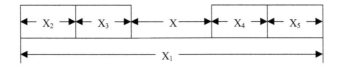

FIGURE 2.10
Assembly in Example 2.6.

using Eq. (2.30) and setting all $g_i = 6.0$ for the normal distribution:

$$t = \sqrt{t_1^2 + t_2^2 + \cdots + t_k^2} \qquad (2.54)$$

Let the total spread of X (that is, the range of all *possible* values of X) be T_s. From Figure 2.10:

$$T_s = 2\sum_{i=1}^{k} a_i + \sqrt{\sum_{i=1}^{k} t_i^2} \qquad (2.55)$$

As $a_i = \dfrac{(T_i - t_i)}{2}$ (Eqs. (1.48) and (1.50)) and $2a_i = (T_i - t_i)$, Eq. (2.53) can be written as:

$$T_s = \sum_{i=1}^{k}(T_i - t_i) + \sqrt{\sum_{i=1}^{k} t_i^2}$$

$$= \sum_{i=1}^{k} T_i - \sum_{i=1}^{k} t_i + \sqrt{\sum_{i=1}^{k} t_i^2}$$

$$= T_a - \sum_{i=1}^{k} t_i + \sqrt{\sum_{i=1}^{k} t_i^2} \qquad (2.56)$$

where T_a is the assembly tolerance obtained by adding all the component tolerances (Eq. (2.26) with $T_{a_i} = T_i$).

When the processes are able to satisfy the tolerances exactly and when $\mu_i = B_i$ for all i,

$$T_i = t_i \text{ and } a_i = 0, \quad \text{for } i = 1, 2, ..., k$$

Hence,

$$T_s = \sqrt{\sum_{i=1}^{k} t_i^2} = T_p$$

which is the probabilistic relationship as per Eq. (2.27).

Rewriting Eq. (2.54):

$$T_a = T_s + \sum_{i=1}^{k} t_i - \sqrt{\sum_{i=1}^{k} t_i^2} \tag{2.57}$$

In the above equation, the t_i is known if the processes have been selected for the manufacture of components (each t_i is the natural process tolerance). Let $\frac{T_i}{t_i} = G$ (a constant), for $i = 1, 2, ..., k$. This ratio is called the *process capability ratio* (PCR, or C_p), and will be discussed in Chapter 4. For a process generating a quality characteristic that follows a normal distribution, t_i is usually taken as $6\sigma_i$. Some industries select processes for manufacturing such that $T_i = 12\sigma_i$, so G for such industries is

$$G = \frac{12\sigma_i}{6\sigma_i} = 2(C_p)$$

It is reasonable to assume that G is the same for all quality characteristics, hence:

$$\frac{T_1}{t_1} = \frac{T_2}{t_2} = \cdots = \frac{T_k}{t_k} = G$$

from which

$$T_i = Gt_i \tag{2.58}$$

From Eq. (2.26), setting $T_{a_i} = T_i$,

$$T_a = T_1 + T_2 + \cdots + T_k$$

which, using Eq. (2.56), becomes:

$$T_a = Gt_1 + Gt_2 + \cdots + Gt_k$$

$$= G\left(\sum_{i=1}^{k} t_i\right) \tag{2.59}$$

Combining Eqs. (2.55) and (2.57), we get:

$$G\left(\sum_{i=1}^{k} t_i\right) = T_s + \sum_{i=1}^{k} t_i - \sqrt{\sum_{i=1}^{k} t_i^2}$$

from which G can be written as:

$$G = \frac{T_s + \Sigma_{i=1}^{k} t_i - \sqrt{\Sigma_{i=1}^{k} t_i^2}}{\Sigma_{i=1}^{k} t_i} \tag{2.60}$$

In the above equation, G (PCR, or C_p) is specified by the manufacturer (as a target value), and T_s can be set equal to T, which is the given (known) assembly tolerance. So the only unknown variables are t_i, $i = 1,\dots, k$. Once these are found, the tolerances $T_1,\dots, T_k$ can be obtained using Eq. (2.56). But, the main problem in solving for t_i, $i = 1, 2,\dots, k$, is the nonlinear term in Eq. (2.58). The following approximate solution procedure can be used to obtain the tolerances, T_i.

Approximate Solution Procedure

If we assume that all the natural process tolerances are equal—that is, $t_1 = t_2 = \cdots = t_k = t$—then:

$$\sqrt{\sum_{i=1}^{k} t_i^2} = \sqrt{kt^2} = t\sqrt{k}$$

$$= \frac{kt}{\sqrt{k}} = \frac{\Sigma_{i=1}^{k} t_i}{\sqrt{k}}$$

So,

$$G = \frac{T_s + \Sigma_{i=1}^{k} t_i - \dfrac{\Sigma_{i=1}^{k} t_i}{\sqrt{k}}}{\Sigma_{i=1}^{k} t_i} \tag{2.61}$$

from which,

$$\sum_{i=1}^{k} t_i = \frac{T_s}{G-1+\frac{1}{\sqrt{k}}} \tag{2.62}$$

Now, t_i and then T_i can be obtained using the following steps:

1. Specify a value for G.
2. Set $T_s = T$ in Eq. (2.44) and find $\sum_{i=1}^{k} t_i$.
3. Based on the natural process tolerances of the available processes on the shop floor, select the processes (that is, select each t_i) such that $\sum_{i=1}^{k} t_i$ is as close as possible to the value obtained in step 2.
4. Find the actual value of G using Eq. (2.58).
5. Depending upon the difference between the specified G and the actual G, change the t_i if necessary.
6. Find T_i for $i = 1, 2,\dots, k$, using Eq. (2.56).

Example 2.6
An assembly consists of five components. The tolerance on the assembly characteristic is given as 0.001. The manufacturer of the components wants the minimum value of G to be 1.33. The natural process tolerances (process capabilities) of the processes available on the shop floor for manufacturing the components are given in Table 2.4. Select the suitable processes and find the tolerances, $T_i, i = 1, 2,\dots, k$.

SOLUTION
Given $k = 5$; $T = 0.001$; $G = 1.33$:

Step 1—The value of G is given as 1.33.
Step 2—Setting $T_s = T = 0.001$ and $G = 1.33$ in (1.60) gives

$$\sum_{i=1}^{5} t_i = \frac{0.001}{1.33-1+1/\sqrt{5}}$$
$$= 0.00129$$

TABLE 2.4

Available Processes (t_i Values)

$i = 1$	$i = 2$	$i = 3$	$i = 4$	$i = 5$
A_1 0.0002	A_2 0.0001	A_3 0.0001	A_4 0.0002	A_5 0.0001
B_1 0.0003	B_2 0.0002	B_3 0.0003	B_4 0.0004	B_5 0.0004
C_1 0.0004	C_2 0.0003	C_3 0.0005	C_4 0.0006	C_5 0.0005
D_1 0.0005	D_2 0.0004	D_3 0.0006		
E_1 0.0006	E_2 0.0005	E_3 0.0007		

Step 3—One possible combination is to select:

	t_i
A_1	0.0002
C_2	0.0003
B_3	0.0003
B_4	0.0004
A_5	0.0001
$\sum t_i =$	0.0013

Step 4—The actual value of G using Eq. (2.42) is obtained as:

$$G = \frac{0.001 + 0.00130 - \sqrt{0.0002^2 + \cdots + 0.0001^2}}{0.00130}$$

$$= 1.29.$$

Step 5—The specified value of G is 1.33. Assume that the manufacturer is not satisfied with the value of G obtained. Let us select the following combination of processes:

	t_i
A_1	0.0002
B_2	0.0002
B_3	0.0003
B_4	0.0004
A_5	0.0001
$\sum t_i =$	0.0012

Using (2.58), now G is 1.35. This is larger than the specified value, thus the manufacturer decides to use these processes, which are A_1, B_2, B_3, B_4, and A_5.

Step 6—The tolerance, T_i, is now obtained using Eq. (2.56) as follows:

$$
\begin{aligned}
T_1 &= t_1 \times 1.35 = 0.0002 \times 1.35 = 0.00027 & (0.0003) \\
T_2 &= t_2 \times 1.35 = 0.0002 \times 1.35 = 0.00027 & (0.0003) \\
T_3 &= t_3 \times 1.35 = 0.0003 \times 1.35 = 0.000405 & (0.0004) \\
T_4 &= t_4 \times 1.35 = 0.0004 \times 1.35 = 0.00054 & (0.0005) \\
T_5 &= t_5 \times 1.35 = 0.0001 \times 1.35 = 0.000135 & (0.0002) \\
& \qquad\qquad\qquad\qquad \sum T_i = 0.0017
\end{aligned}
$$

(The values within parentheses are rounded off to four significant digits after the decimal point.)

It can be seen that the actual range of the assembly characteristic is $\frac{(0.0017 - 0.001)}{0.001} \times 100 = 73\%$ larger than the allowable range, which is $T = 0.001$.

We will now represent this increase in the range in the probability distribution of the assembly characteristic. First, we will compute the actual variance of the assembly characteristic, X. For $X = X_1 \pm X_2 \pm X_3 \pm X_4 \pm X_5$,

$$\sigma^2 = \sqrt{\sigma_1^2 + \sigma_2^2 + \cdots + \sigma_5^2}$$

$$\sigma_1 = \frac{t_1}{6} = \frac{0.0002}{6} = 0.000033$$

$$\sigma_2 = \frac{t_2}{6} = \frac{0.0002}{6} = 0.000033$$

$$\sigma_3 = \frac{t_3}{6} = \frac{0.0003}{6} = 0.000050$$

$$\sigma_4 = \frac{t_4}{6} = \frac{0.0004}{6} = 0.000067$$

$$\sigma_5 = \frac{t_5}{6} = \frac{0.0001}{6} = 0.000017$$

$$\sigma^2 = (0.000033)^2 + \cdots + (0.000017)^2$$

$$= 9455999 \times 10^{-9}; \; \sigma = 0.000097$$

As X is normally distributed, the natural process tolerance for X (the range containing 99.73% of all values) at any given location of its mean is

$$t = 6\sigma = 6 \times 0.000097 = 0.000582$$

Now let us calculate the a_i values for $i = 1, 2, \ldots, k$:

$$a_1 = \frac{(T_1 - t_1)}{2} = \frac{(0.0003 - 0.0002)}{2} = 0.00005$$

$$a_2 = \frac{(0.0003 - 0.0002)}{2} = 0.00005$$

$$a_3 = a_4 = a_5 = 0.00005$$

The distribution of X, with these values, is given in Figure 2.11.

The formulas derived and the values we obtained assume the worst-case scenario (that is, the shift between μ_i and B_i is either $+a_i$ or $-a_i$ for all i), but in most of the situations, the shift between μ_i and B_i may not be equal to a_i for all i. Only after the manufacture is completed will we know the exact location of μ_i for all i and hence the location of the mean of X, which is μ. So, we do not have to consider an interval or range for μ.

Let us consider a case in which the maximum shift allowed (that is, a_i) is *not* used while setting the processes. We will use the data from Example 2.6.

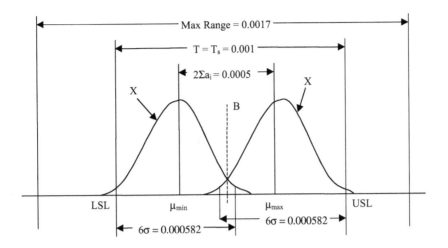

FIGURE 2.11
Expected distribution of X in Example 2.6.

The maximum shifts (a_i) and the actual shifts are given below.

Component (i)	$a_i = (T_i - t_i)/2$	Actual Shift
1	$(0.0003 - 0.0002)/2 = 0.00005$	+0.00002
2	$(0.0003 - 0.0002)/2 = 0.00005$	−0.00001
3	$(0.0004 - 0.0003)/2 = 0.00005$	+0.00004
4	0.00005	+0.00005
5	0.00005	−0.00002

Let us assume that the assembly is as shown in Figure 2.10. In this assembly, the assembly characteristic is

$$X = X_1 - X_2 - X_3 - X_4 - X_5$$

The nominal size of X is

$$B = B_1 - B_2 - B_3 - B_4 - B_5$$

and the mean of X is

$$\mu = \mu_1 - \mu_2 - \mu_3 - \mu_4 - \mu_5$$

Based on the assumptions about the actual shifts,

$$\mu_1 = B_1 + 0.00002; \quad B_1 = \mu_1 - 0.00002$$
$$\mu_2 = B_2 - 0.00001; \quad B_2 = \mu_2 + 0.00001$$
$$\mu_3 = B_3 + 0.00004; \quad B_3 = \mu_3 - 0.00004$$
$$\mu_4 = B_4 + 0.00005; \quad B_4 = \mu_4 - 0.00005$$
$$\mu_5 = B_5 - 0.00002; \quad B_5 = \mu_5 + 0.00002$$

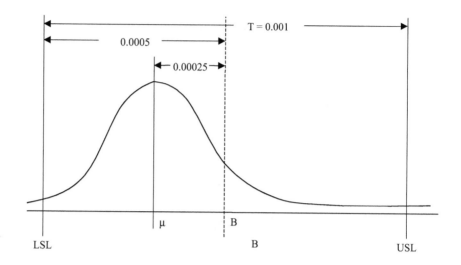

FIGURE 2.12
Actual distribution of X in Example 2.6.

Hence, the nominal size of X is

$$
\begin{aligned}
B &= (\mu_1 - 0.00002) - (\mu_2 + 0.00001) - (\mu_3 - 0.00004) \\
&\quad - (\mu_4 - 0.00005) - (\mu_5 + 0.00002) \\
&= (\mu_1 - \mu_2 - \mu_3 - \mu_4 - \mu_5) - 0.00002 - 0.00001 \\
&\quad + 0.00004 + 0.00005 - 0.00002 \\
&= \mu + 0.00004 \\
\mu &= B - 0.00004
\end{aligned}
$$

That is, the mean of distribution of X is located to the left of the nominal size B at a distance of 0.00004. From earlier calculation, the standard deviation of X is

$$\sigma = 0.000097$$

The actual distribution of X is given in Figure 2.12. Now the actual proportion of undersized and oversized components can be estimated separately.

UNDERSIZED

$$
\begin{aligned}
\mu - \text{LSL} &= 0.0005 - 0.00004 = 0.00046 \\
\sigma &= 0.000097 \\
\frac{(\mu - \text{LSL})}{\sigma} &= \frac{0.00046}{0.000097} = 4.74; \quad (\mu - \text{LSL}) = 4.74\sigma
\end{aligned}
$$

Table A.5 in the Appendix gives the proportion of defectives for some standard normal values not found in the commonly used standard normal tables. It contains the cumulative probabilities for Z values in the range (–3.00 to –6.00) and the parts per million (ppm) values obtained by multiplying the cumulative probabilities by 10^6. From the table, the proportion of undersized components (p_{us}) is $z = -4.74 = 1.07$ ppm.

OVERSIZED

$$(USL - \mu) = 0.0005 + 0.00004 = 0.00054$$

$$= \frac{0.0005}{0.000097} \, \sigma = 5.567\sigma$$

From Table A.5, the proportion of oversized components (p_{os}) is $z = -5.57 = 0.0128$ ppm. In none of the tolerance assignment methods discussed earlier was an objective function used. In the next section, minimization of the total cost of manufacture will be used as the objective function in allocating tolerance among components.

2.6 Tolerance Allocation that Minimizes the Total Manufacturing Cost

(This section is based on the paper by Bennett and Gupta, "Least-Cost Tolerances—I."[2]) The objective here is to determine the component tolerances for a given (specified) assembly tolerance. The main differences between this discussion and the earlier methods we used to allocate tolerances among the components of an assembly are as follows:

1. An objective function is included: to minimize the total cost of manufacturing the components.
2. The relationship between the assembly tolerance, *T*, and the component tolerances, T_i is given by:

$$T = \sum_{i=1}^{k} N_i \times T_i \qquad (2.63)$$

3. We can see that the relationship is still linear, but the coefficients of the T_i need not be equal to 1. (In the paper by Mansoor[6] and earlier discussions, we assumed that $N_i = 1$ for all *i*.)
4. Process selection is not explicitly addressed (t_i is *not* used). It is assumed that the manufacturing cost of component *i* (with characteristic X_i and tolerance T_i) is assumed to be

$$C_i = h_i \times T_i^{\alpha_i} \qquad (2.64)$$

In this equation, h_i and α_i are constants which affect the shape of the cost–tolerance curve, and $\alpha_i < 0$. This cost relationship satisfies two basic requirements: (1) When $T_i = 0$, $C_i = \infty$; and (2) C_i should be a decreasing function of T_i.

2.6.1 Formulation of the Problem

The decision variables are T_1, T_2, ..., T_k. The objective function is to minimize the total cost of manufacture. That is, to minimize:

$$C = \sum_{i=1}^{k} h_i \, T_i^{\alpha_i} \tag{2.65}$$

where C is the total manufacturing cost of the components. The constraint in this problem guarantees that the sum of the component tolerances selected does not exceed the given tolerance on the assembly characteristic; that is,

$$\sum_{i=1}^{k} N_i T_i = T \tag{2.66}$$

All the component tolerances are non-negative; that is,

$$T_i \geq 0, \quad i = 1, 2, ..., k$$

Now the formulation can be summarized as find:

$$T_1, T_2, ..., T_k$$

so as to minimize:

$$C = \sum_{i=1}^{k} h_i T_i^{\alpha_i}$$

subject to:

$$\sum_{i=1}^{k} N_i T_i = T$$

and

$$T_i \geq 0.0, \quad \text{for all } i$$

(T_i's are assumed to be continuous.)

This is a nonlinear programming problem that can be solved using the method of LaGrange multipliers.[1] The first step is to transform this *con-strained* optimization problem to an *unconstrained* optimization problem. The objective function of this unconstrained problem is written as:

$$F = \sum_{i=1}^{k} h_i T_i^{\alpha_i} + \lambda \left[\sum_{i=1}^{k} N_i T_i - T \right] \tag{2.67}$$

Now the problem is to find the optimum values of T_i, $i = 1, 2, ..., k$, and λ that minimize F. This is a simpler problem than the original problem because the optimum values of T_i and λ can be obtained by differentiating F with respect to the T_i and λ and setting the derivatives equal to 0:

$$\frac{\partial F}{\partial T_i} = h_i \alpha_i T_i^{\alpha_i - 1} + \lambda N_i = 0, \quad i = 1, 2, ..., k$$

$$h_i \alpha_i T_i^{\alpha_i - 1} = -\lambda N_i$$

$$T_i = \left[\frac{-\lambda N_i}{h_i \alpha_i} \right]^{\frac{1}{\alpha_i - 1}}, \quad i = 1, 2, ..., k \tag{2.68}$$

$$< 0$$

$$\frac{\partial F}{\partial \lambda} = \sum_{i=1}^{k} N_i T_i - T = 0$$

$$T = \sum_{i=1}^{k} N_i T_i = \sum_{i=1}^{k} N_i \left[\frac{-\lambda N_i}{h_i \alpha_i} \right]^{\frac{1}{\alpha_i - 1}}$$

$$= \sum_{i=1}^{k} N_i \left(\frac{-1}{h_i \alpha_i} \right)^{\frac{1}{\alpha_i - 1}} (\lambda N_i)^{\frac{1}{\alpha_i - 1}} \tag{2.69}$$

Let:

$$K_i = N_i \left(-\frac{1}{\alpha_i h_i} \right)^{\left(\frac{1}{\alpha_i - 1} \right)} \quad \text{and} \quad b_i = \left(\frac{1}{\alpha_i - 1} \right)$$

Hence,

$$T = \sum_{i=1}^{k} K_i (\lambda N_i)^{b_i} \tag{2.70}$$

In the above equation, the only unknown variable is λ. If the value of λ that satisfies the above equation can be found, then Eq. (2.66) can be used to obtain T_i, $i = 1, 2, \ldots, k$.

Let us consider the following equation obtained from Eq. (2.68):

$$f(\lambda) = \sum_{i=1}^{k} K_i(\lambda N_i)^{b_i} - T \tag{2.71}$$

This is a polynomial in λ. The problem now is to find the root of Eq. (2.69), which is equivalent to solving for λ. The following is a modified Newton-Raphson method.[7]

2.6.2 Steps for the Newton–Raphson Method

1. First find $f'(\lambda) = \frac{df(\lambda)}{d\lambda}$ and $f''(\lambda) = \frac{d^2 f(\lambda)}{d\lambda^2}$:

$$f'(\lambda) = \sum_{i=1}^{k} K_i b_i(\lambda N_i)^{(b_i - 1)} N_i \tag{2.72}$$

$$f''(\lambda) = \sum_{i=1}^{k} K_i b_i(b_i - 1)(\lambda N_i)^{(b_i - 2)} N_i^2 \tag{2.73}$$

2. Assume a starting value for $\lambda(\lambda_0)$. A possible value is

$$\lambda_0 = \left[\frac{T}{\sum_{i=1}^{k}(K_i N_i^{b_i})} \right]^{(\bar{\alpha} - 1)} \tag{2.74}$$

where

$$\bar{\alpha} = \frac{(\sum_{i=1}^{k} N_i \alpha_i)}{(\sum_{i=1}^{k} N_i)} \tag{2.75}$$

Set $\lambda = \lambda_0$.

3. Find:

$$U(\lambda) = \frac{f(\lambda)}{f'(\lambda)} \tag{2.76}$$

and

$$U'(\lambda) = 1 - \frac{f(\lambda)f''(\lambda)}{[f'(\lambda)]^2} \tag{2.77}$$

4. Find:

$$\delta = -\frac{U(\lambda)}{U'(\lambda)} \tag{2.78}$$

5. Set $\lambda = \lambda + \delta$.
6. If $|\delta| <$ some pre-specified value ε, stop. λ is the root; otherwise, go to step 3.

Example 2.7

An assembly consisting of six components is considered. The tolerance on the assembly characteristic is 0.012. The α_i and h_i values of the six components are as follows:

Component i	α_i	h_i
A	−0.60	1.00
B	−0.80	1.75
C	−0.60	0.90
D	−0.40	1.30
E	−0.40	1.10
G	−0.70	0.95
$N_i = 1$ for all i.		

Find $T_A, T_B, \ldots, T_G$.

SOLUTION
Step 1:

$$f(\lambda) = \sum_{i=1}^{6} K_i (\lambda N_i)^{b_i} - T$$

$$K_1 = \left(-\frac{1}{(-0.60) \times (1.00)} \right)^{\frac{1}{(-0.6-1)}} = 0.72668$$

$$K_2 = \left(-\frac{1}{(-0.80) \times (1.75)} \right)^{\frac{1}{(-0.8-1)}} = 1.205542$$

$$K_3 = 0.6803715; \quad K_4 = 0.6268235$$

$$K_5 = 0.5563183; \quad K_6 = 0.7866426$$

$$b_1 = \frac{1}{-0.6 - 1} = -0.625$$

$$b_2 = \frac{1}{-0.8 - 1} = -0.5556$$

$$b_3 = -0.625; \quad b_4 = -0.7143$$

$$b_5 = -0.7143; \quad b_6 = -0.5882$$

$$f(\lambda) = 0.72668(\lambda)^{-0.625} + 1.205542(\lambda)^{-0.5556} + 0.6803715(\lambda)^{-0.625}$$
$$+ 0.626823(\lambda)^{-0.7143} + 0.5563183(\lambda)^{-0.7143} + 0.7866426(\lambda)^{-0.5882} - 0.012$$

The objective is to find the value of λ that makes $f(\lambda) = 0.0$. The function $f(\lambda)$ is a monotonically decreasing function of λ, because each term with λ in $f(\lambda)$ is a monotonically decreasing function of λ. This implies that $f(\lambda)$ has at most one root. If the optimization problem has a feasible solution, then there has to be at least one value of λ that satisfies $f(\lambda) = 0$. Hence, $f(\lambda)$ has exactly one root.

Now the modified Newton–Raphson method is used to find this root. After finding the first and second derivatives of $f(\lambda)$, the following steps are executed.

Step 2:

$$\lambda_0 = \left[\frac{0.012}{\Sigma_{i=1}^6 K_i}\right]^{(\bar{\alpha}-1)} = 12{,}246.83$$

$$\left(\bar{\alpha} = \frac{\Sigma_{i=1}^6 \alpha_i}{6} = -0.5833\right)$$

as $N_i = 1$, for all i.
   Set $\lambda = \lambda_0$.
Steps 3, 4, 5, and 6:

$$f'(\lambda) = \sum_{i=1}^6 K_i b_i (\lambda)^{b_i - 1}$$

$$f''(\lambda) = \sum_{i=1}^6 K_i b_i (b_i - 1)(\lambda)^{b_i - 2}$$

Iteration	λ	δ
1	12,246.83	8368.66
2	20,615.49	−2693.33
3	17,922.16	−306.04
4	17,616.12	−4.28
5	17,611.84	0
6	17,611.84	0
λ = 17,611.84		

Let us solve this problem with different starting values for λ.

TRIAL 1: λ_0 = 800,000

Iteration	λ	δ
1	800,000	−482,170.20
2	317,829.80	−184,580.60
...		
8	17,622.01	−10.12
9	17,611.89	−0.00919
10	17,611.88	0
11	17,611.88	0
λ = 17,611.88		

TRIAL 2: λ_0 = 8000

Iteration	λ	δ
1	8000	−811,748.90

The values of λ moved away from 17,611.88 and the solution did not converge. In this problem, the starting value for λ has to be at least 8050 (approximately) in order for the solution to converge to the correct value. Hence, it is advised that the starting value recommended earlier be used.

Once the value of λ has been obtained, the values of T_A, T_B, ..., and T_G can be determined using Eq. (2.66):

$$T_A = \left[\frac{-17611.84}{(1.00)(-0.60)} \right]^{\frac{1}{(-0.6-1)}} = 0.0016133 \ (0.0016)$$

The value for T_A is rounded off to 0.0016.

$$T_B = \left[\frac{-17611.84}{(1.75)(-0.80)} \right]^{\frac{1}{(-0.8-1)}} = 0.0052771 \ (0.0053)$$

$$T_C = 0.0015105 \quad (0.0015)$$
$$T_D = 0.00058133 \ (0.0006)$$
$$T_E = 0.00051594 \ (0.0005)$$
$$T_G = 0.0025017 \quad (0.0025)$$

Total: 0.0120

The final component tolerances and the associated manufacturing costs are as follows:

	Final T_i	C_i (Cost of Manufacture) ($)
A	0.0016	47.59
B	0.0053	115.78
C	0.0015	44.52
D	0.0005	27.19
E	0.0005	23.00
G	0.0025	62.97
Total	0.0120	321.05

In the next section, this method is extended to problems with more than one assembly characteristic.

2.7 Tolerance Allocation in Assemblies with More Than One Quality Characteristic

This section is based on the paper, "Least-Cost Tolerances—II."[3] Let us assume that an assembly, consisting of five components, has two quality characteristics, X and Y. Let the relationships among the characteristics be as follows:

$$X = X_1 + X_2 + X_3 + X_4 + X_5$$
$$Y = X_2 + X_3 - X_5$$

So,

$$T_x = T_1 + T_2 + T_3 + T_4 + T_5$$
$$T_y = T_2 + T_3 + T_5$$

In general, let us assume that an assembly has m characteristics with the following relationships among the tolerances:

$$T^{(1)} = \sum_{i=1}^{k} N_{1i} T_i$$

$$T^{(2)} = \sum_{i=1}^{k} N_{2i} T_i$$

$$T^{(m)} = \sum_{i=1}^{k} N_{mi} T_i \qquad (2.79)$$

Each of the above m relationships is a constraint.

The total manufacturing cost that has to be minimized is

$$C = \sum_{i=1}^{k} h_i T_i^{\alpha_i}$$

Now the problem can be stated as: Find T_1, T_2, ..., and T_k so as to minimize $C = \sum_{i=1}^{k} h_i T_i^{\alpha_i}$, subject to $\sum_{i=1}^{k} N_{ji} T_i = T^{(j)}$, for $j = 1, 2, ..., m$, and $T_i > 0$, for all i.

This problem can be solved using LaGrange multipliers, with one LaGrange multiplier for each constraint. The objective function for the unconstrained optimization problem is

$$F = \sum_{i=1}^{k} h_i T_i^{\alpha_i} + \lambda_1 \left[\sum_{i=1}^{k} N_{1i} T_i - T^{(1)} \right] + \cdots + \lambda_2 \left[\sum_{i=1}^{k} N_{2i} T_i - T^{(2)} \right] \cdots$$

$$+ \lambda_m \left[\sum_{i=1}^{k} N_{mi} T_i - T^{(m)} \right] \qquad (2.80)$$

The values of T_1, T_2, ..., T_k and λ_1, λ_2, ..., and λ_m [$(k + m)$ decision variables] can be obtained by solving the following equations:

$$\frac{dF}{dT_i} = 0, \quad \text{for } i = 1, 2, ..., k$$

and

$$\frac{dF}{d\lambda_j} = 0, \quad \text{for } j = 1, 2, ..., m$$

2.8 Tolerance Allocation When the Number of Processes is Finite

The formulation of the tolerance problem in the previous section assumes that the decision variables T_1, T_2, ..., T_k are continuous, which indirectly assumes that there is an infinite number of processes available. This may not be realistic in real-life applications. Let us now consider the situation for which there is only a finite number of processes available.

2.8.1 Assumptions

1. The assembly has k components.
2. There are l_i processes available for machining component i (for generating characteristic X_i), $i = 1, 2, ..., k$.

3. The cost per unit time of process j for part i is C_{ij}, $i = 1, 2, ..., k$ and $j = 1, 2, ..., l_i$.

4. The relationship among the tolerances is assumed to be

$$T = \sum_{i=1}^{k} N_i T_i$$

5. The process capability ratio for all the processes is G; that is,

$$\frac{T_i}{t_{ij}} = G, \quad \text{for all } i \text{ and } j \tag{2.81}$$

where t_{ij} is the natural process tolerance of process j used for component i, $i = 1, 2, ..., k$ and $j = 1, 2, ..., l_i$.

6. The unit cost of machining component i on process j is C_{ij}, $i = 1, 2, ...,$ k and $j = 1, 2, ..., l_i$.

The problem is to find $T_1, T_2, ..., T_k$ that minimize the total cost of manufacture and satisfy all constraints. This problem can be formulated as a mathematical programming problem using zero–one decision variables.[5]

2.8.2 Decision Variables

$$\begin{aligned} Y_{ij} &= 1 \text{ (yes)}, \quad \text{if process } j \text{ is selected for component } i \\ &= 0 \text{ (no)}, \quad \text{if process } j \text{ is not selected for component } i \end{aligned}$$

2.8.3 Objective Function

Minimize

$$C = \sum_{i=1}^{k} \sum_{j=1}^{l_i} C_{ij} Y_{ij}$$

2.8.4 Constraints

The first constraint guarantees that the sum of the tolerances of components does not exceed the given tolerance on the assembly characteristic:

I.

$$\sum_{i=1}^{k} T_i \leq T \rightarrow \sum_{i=1}^{k} \sum_{i=1}^{l_i} G t_{i,j} Y_{i,j} \leq T$$

$$\sum_{i=1}^{k} \sum_{i=1}^{l_i} t_{i,j} Y_{i,j} \leq \frac{T}{G} \quad \text{(1 constraint)} \tag{2.82}$$

The second constraint set ensures that exactly one process is selected for each of the k components.

II.

$$\sum_{j=1}^{l_i} Y_{ij} = 1, \quad \text{for } i = 1, 2, 3, \ldots, k \quad (k \text{ constraints}) \tag{2.83}$$

This is a zero–one integer programming problem and can be solved by a linear programming software such as LINDO.

Example 2.8

An assembly consists of three components. The relationship among the quality characteristics is as follows:

$$X = X_1 - X_2 + X_3$$

The processes available for processing the components along with the respective unit costs and the natural process tolerances are as follows:

Process	t	Unit Cost ($)
Component 1		
1	0.001	50
2	0.002	35
3	0.005	25
Component 2		
1	0.001	60
2	0.003	35
Component 3		
1	0.004	30
2	0.006	18

The tolerance on the assembly characteristic is 0.011, and the ratio of the component tolerance to the natural process tolerance is 1.3 (G) for all components. Formulate this as a zero–one integer programming problem to find T_1, for $i = 1, 2,$ and 3 to minimize the total cost of manufacture.

2.8.5   Formulation

2.8.5.1   *Decision Variable*

Y_{ij} = 1,   if process j is selected for manufacturing component i

     = 0,   if process j is not selected for manufacturing component i

2.8.5.2   Objective Function

Minimize

$$C = \sum_{i=1}^{k}\sum_{j=1}^{l_i} C_{ij}Y_{ij}$$

$$= 50\ Y_{11} + 35\ Y_{12} + 25\ Y_{13} + 60\ Y_{21} + 35\ Y_{22} + 30\ Y_{31} + 18\ Y_{32}$$

2.8.5.3   Constraints

1. $\sum_{i=1}^{k}\sum_{j=1}^{l_i} t_{ij}Y_{ij} \le \dfrac{T}{G}$

$$0.001\ Y_{11} + 0.002\ Y_{12} + 0.005\ Y_{13} + 0.001\ Y_{21} + 0.003\ Y_{22}$$
$$+ 0.004\ Y_{31} + 0.006\ Y_{32} \le 0.011/1.3(0.0085)$$

2. $\sum_{j=1}^{l_i} Y_{ij} = 1,\quad$ for $i = 1, 2, 3, \ldots, k$ (3 constraints)

$$Y_{11} + Y_{12} + Y_{13} = 1.0$$
$$Y_{21} + Y_{22} = 1.0$$
$$Y_{31} + Y_{32} = 1.0$$

2.8.5.4   Solution

This problem was solved using LINDO, and the solution was obtained as follows:

$Y_{12} = 1.0\ (Y_{11} = Y_{13} = 0)$; process 2 is selected for component 1.
$Y_{21} = 1.0\ (Y_{22} = 0)$; process 1 is selected for component 2.
$Y_{31} = 1.0\ (Y_{32} = 0)$; process 1 is selected for component 3.

Now, the tolerances are determined as follows:

$t_{12} = 0.002$; hence, $T_1 = 0.002 \times 1.3 = 0.0036$.
$t_{21} = 0.001$; hence, $T_2 = 0.001 \times 1.3 = 0.0013$.
$t_{31} = 0.004$; hence, $T_3 = 0.004 \times 1.3 = 0.0052$.

The sum of T_i's is 0.0101.

It may be noted that this example problem, being small, can be solved by hand and does not require a formulation and software to solve. The advantage of formulation and software is obvious in large-scale problems.

In all the tolerance allocation problems we have studied so far, the following linear relationship among the tolerances was assumed:

$$X = X_1 \pm X_2 \pm X_3 \pm \cdots \pm X_5$$

This assumption about the linear relationship will be relaxed in the next section.

2.9 Tolerance Allocation for Nonlinear Relationships among Components

So far, we assumed that $N_i = 1$ for all i in the equation:

$$\sum_{i=1}^{k} N_i T_i = T$$

N_i represents the contribution of dimension X_i (characteristic X_i) to the assembly characteristic X. This is also the sensitivity of assembly characteristic X to characteristic X_i and is equal to the ratio of the change in the assembly characteristic X to a small change in X_i. That is, if $X = f(X_1, X_2, ..., X_k)$, then:

$$N_i = \left| \frac{\partial f(X_1, X_2, ..., X_k)}{\partial X_k} \right|, \quad i = 1, 2, ..., k \qquad (2.84)$$

which can be used in the relation:

$$\sum_{i=1}^{k} N_i T_i = T$$

Example 2.9

Consider the simple helical spring shown in Figure 2.13. There are two assembly characteristics of interest to the user:

1. Spring rate (R), which is equal to:

$$R = \frac{E \times (d_w)^4}{8(d_i + d_w)^3 \times M} \quad (\text{lb/in.}) \qquad (2.85)$$

   where E = modulus of elasticity in shear; d_w = wire diameter; d_i = inside diameter of spring; and M = number of active coils.

2. Outside diameter of the spring (d_o), which is equal to:

$$d_0 = d_i + 2d_w \qquad (2.86)$$

We have to find $(dR)/(d(d_w))| = N_{d_w}^R$, the contribution of d_w to R; $(dR)/(d(d_j)) = N_{d_i}^R$, the contribution of d_i to R; $(dR)/(dM) = N_M^R$, the contribution of M to R; $(d(d_0))/(d(d_i)) = N_{d_i}^{d_0}$, the contribution of d_i to d_0; $(d(d_0))/(d(d_w)) = N_{d_w}^{d_0}$, the contribution of d_w to d_0.

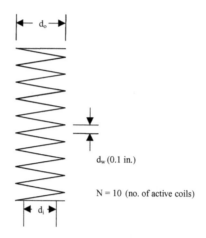

FIGURE 2.13
Spring in Example 2.9.

Let the nominal sizes of d_w, d_i, and M be 0.1", 1.0", and 10", respectively, and let $E = 11.5 \times 10^6$:

$$\frac{dR}{d(d_w)} = \frac{E}{8M}\left[\frac{4d_w^3}{(d_i + d_w)^3} - \frac{3d_w^4}{(d_i + d_w)^4}\right]$$

$$= \frac{10.5 \times 10^6}{8 \times 10}\left[\frac{4 \times 0.1^3}{(1 + 0.1)^3} - \frac{3 \times 0.1^4}{(1 + 0.1)^4}\right] = 402.50$$

That is, $N_{d_w}^R = 402.50$.

$$\frac{dR}{dM} = -\frac{Ed_w^4}{8(d_i + d_w)^3} \times \frac{1}{M^2}$$

$$= \frac{11.5 \times 10^6 \times 0.1^4}{8 \times (1.1)^3} \times \frac{1}{100}$$

$$= -1.08$$

That is, $N_M^R = 1.08$.

$$\frac{d(d_o)}{d(d_i)} = 1 = N_{d_i}^{d_o}$$

$$\frac{d(d_o)}{d(d_w)} = 2 = N_{d_w}^{d_o}$$

Let the specified tolerance on R be $B_R \pm 0.54$ lb/in. ($T_R = 2 \times 0.54 = 1.08$) and the tolerance on d_o be $B_{d_o} \pm 0.020$" ($T_{d_o} = 2 \times 0.020 = 0.04$").

The problem is to find the optimum tolerances on the component characteristics: d_w, d_i, and M, so that the total cost of manufacture is minimized. The
constraints are

$$402.5\ T_1 + 37.99\ T_2 + 1.08\ T_3 \leq 1.08 \qquad (2.87)$$

and

$$2\ T_1 + T_2 \leq 0.04 \qquad (2.88)$$

The constraint in Eq. (2.84) relates the tolerances on d_w, d_i, and M to the given
tolerance on R and the constraint in Eq. (2.85) relates the tolerances on d_w and
d_i to the given tolerance on d_o.
    A suitable objective function such as the cost minimization objective func-
tion given earlier in Eq. (2.63) can be used in this problem; that is, minimize:

$$C = \sum_{i=1}^{3} h_i T_i^{\alpha_i} \qquad (2.89)$$

Using the LaGrange multiplier, the unconstrained objective function becomes:

$$F = \sum_{i=1}^{3} h_i T_i^{\alpha_i} + \lambda_1 [402.5\ T_1 + 37.99\ T_2 + 1.08\ T_3 - 1.08]$$
$$+ \lambda_2 [2\ T_1 + T_2 - 0.04] \qquad (2.90)$$

Now, the unknowns, T_1, T_2, T_3, λ_1, and λ_2, can be found by differentiating
Eq. (2.87) with respect to these variables and setting the resulting derivatives
to zero. This problem can also be solved using nonlinear programming software such as GINO.

2.10   Other Topics in Tolerancing

This chapter does not cover all the methods for solving tolerance allocation
problems. Dynamic programming can be effectively used to allocate the
assembly tolerance among the components. Zhang and Huq[10] give an excellent review of tolerancing techniques. A recent problem is sequential tolerance control, suitable for parts moving through a sequence of operations
specified in the process plan.[4,8] It uses real-time measurements to sequentially
and selectively adjust the target point of machining operations to maximize
the final output quality. Wheeler et al.[8] developed a probabilistic approach to
select an optimum subset of technological processes required to execute a
process plan under a conventional tolerance control strategy.

2.11 References

1. Bazarra, M.S., Sherali, H.D., and Shetty, C.M., *Nonlinear Programming: Theory and Algorithms*, 2nd ed., John Wiley & Sons, New York, 1993.
2. Bennett, G. and Gupta, L.C., Least-cost tolerances—I, *Int. J. Prod. Res.*, 8(1), 65, 1969.
3. Bennett, G. and Gupta, L.C., Least-cost tolerances—II, *Int. J. Prod. Res.*, 8(2), 169, 1970.
4. Fracticelli, B.P., Lehtihet, E.A., and Cavalier, T.M., Sequential tolerance control in discrete parts manufacturing, *Int. J. Prod. Res.*, 35, 1305, 1997.
5. Ignozio, J.P. and Cavalier, T.M., *Linear Programming*, Prentice Hall, New York, 1994.
6. Mansoor, E.M., The application of probability to tolerances used in engineering designs, *Proc. Inst. Mech. Eng.*, 178(1), 29, 1963–1964.
7. Hornbeck, R.W., *Numerical Methods*, Quantum Publishers, New York, 1975.
8. Wheeler, D.L., Cavalier, T.M., and Lehtihet, E.A., An implicit enumeration approach to tolerance allocation in sequential tolerance control, *IIE Trans.*, 31, 75, 1999.
9. Wheeler, D.L., Cavalier, T.M., and Lehtihet, E.A., An implicit enumeration approach to probabilistic tolerance allocation under conventional tolerance control, *Int. J. Prod. Res.*, 37, 3773, 1999.
10. Zhang, H.C. and Huq, M.E., Tolerancing techniques: the state-of-the art, *Int. J. Prod. Res.*, 30, 2111, 1992.

2.12 Problems

1. In a shaft/sleeve assembly, the inside diameter of the sleeve (X_1) follows normal distribution with a mean of 20 and a standard deviation of 0.3. The outside diameter of the shaft (X_2) follows normal distribution with a mean of 19.6 and a standard deviation of 0.4. The specification for the clearance between the mating parts is 0.5 ± 0.40.

   a. What fraction of assemblies will fail to meet the specification limits?

   b. Ignore the values of the means and standard deviations of X_1 and X_2 given above. Find T_1 and T_2 (tolerances of X_1 and X_2) using the additive and probabilistic relationships, assuming $T_1 = 1.5 \, T_2$.

2. Consider an assembly consisting of three components. The length of the assembly (X) is the sum of the lengths of the three components, X_1, X_2, and X_3, which are normally distributed with means 1.00, 3.00, and 2.00, respectively. The proportions of X, X_1, X_2, and X_3 outside the respective specification limits are specified as 0.0027. Assume that the variances of X_1, X_2, and X_3 are equal and that the means of X_1, X_2, and X_3 are equal to the respective nominal sizes.

a. What are the specification limits for X_1, X_2, and X_3?

b. What are the tolerances for X_1, X_2, and X_3, if additive relationship is used? Assume that $T_1 = T_2 = T_3$, and that the specification limits for the assembly characteristics are 6 ± 0.006.

3. Three resistors are to be connected in series so that their resistances add together for the total resistance. One is a 150-ohm resistor and the other two are 100-ohm resistors. The respective specification limits are 150 ± 7.5 and 100 ± 6. If the resistances of the three resistors are normally distributed and the tolerance for the resistance of each resistor is equal to $6 \times$ the respective standard deviation,

a. What are the two values within which the total resistance will lie 99.73% of the time?

b. How many standard deviations from the nominal value will the limits of the additive tolerance for the total resistance lie?

4. Consider an assembly characteristic $X = X_1 + X_2 - X_3$. The specification (tolerance) limits on X are 3.00 ± 0.05 in. Let the characteristics X_1, X_2, and X_3 be normally and independently distributed with means $\mu_1 = 0.5$ in., $\mu_2 = 1.50$ in., and $\mu_3 = 1.00$ in., respectively. The variances of the characteristics are equal. The processes are selected such that 99.73% of the characteristics fall within the tolerance limits. The means of the characteristics are equal to the respective basic sizes. It is desired that 99.9937% of the assembly characteristic falls within the tolerance limits 3.00 ± 0.05 in. Determine the tolerances (T_i, $i = 1$, 2, and 3) on the characteristics X_1, X_2, and X_3 using the probabilistic tolerance relationship.

5. Consider an assembly such that its characteristic X is $X = X_1 + X_2$, where X_1 and X_2 are the characteristics of the components. The characteristics X_1 and X_2 follow uniform distribution within the respective lower and upper tolerance limits. Assume that the tolerance limits of X_1 and X_2 are equal. The assembly tolerance (T) is equal to 0.002. Find the tolerances on X_1 and X_2 such that 100% of the values of the assembly characteristic X is contained within its tolerance limits.

6. An assembly consists of three components. The relationship between the assembly characteristic X and the component characteristics is $X = X_1 - X_2 - X_3$. The processes available for machining these component characteristics and their natural process tolerances are as follows:

Component 1		Component 2		Component 3	
Process	t	Process	t	Process	t
1	0.0001	1	0.0001	1	0.0004
2	0.0002	2	0.0002	2	0.0005
3	0.0004	3	0.0004	3	0.0006

The tolerance on the assembly dimension is 0.0008. Find the tolerances on X_1, X_2, and X_3. Assume that $G = 1.3$ and that the component characteristics are normally distributed. Use the technique developed by Mansoor. Please try to achieve a value of G which is as close as possible to 1.3, but not less than 1.3.

7. In the above problem, assume that the actual shifts between the means and the respective basic sizes of the component characteristics are as follows:

   Component 1: Actual shift $= -\frac{1}{2}$ of the maximum shift allowed (mean is located to the left of the basic size).

   Component 2: Actual shift $= +$ maximum shift allowed (mean is located to the right of the basic size).

   Component 3: Actual shift $= + \frac{1}{2}$ of the maximum shift allowed.

   Estimate the proportion of undersized and oversized assemblies.

8. An assembly consists of three components. The relationship among the quality characteristics is: $X = X_1 + X_2 - X_3$. The processes for generating these characteristics with the respective unit costs and natural process tolerances are given below:

Component 1			Component 2			Component 3		
Process	t	Unit Cost	Process	t	Unit Cost	Process	t	Unit Cost
1	0.0001	$15	1	0.0001	$13	1	0.0004	$15
2	0.0002	$7	2	0.0002	$8	2	0.0005	$11
3	0.0004	$5	3	0.0004	$6	3	0.0006	$9

   The tolerance on the assembly characteristic is 0.0008. Formulate (that is, define the decision variables and write the objective function and constraints) this problem as a zero–one integer programming problem to find T_i, $i = 1, 2,$ and 3, that minimize the total unit cost of manufacture and achieve a G value that is at least 1.3. (Do not solve for T_i).

9. An assembly consists of five components. The manufacturing costs of these components are given by $C_i = h_i T_i^{\alpha_i}$ where T_i is the tolerance on the characteristic of component i and α_i and h_i are given in the following table:

i	α_i	h_i
1	-0.50	1.50
2	-0.80	1.00
3	-0.70	1.80
4	-0.60	0.90
5	-0.40	1.20

The assembly tolerance is specified as 0.01 and is related to the component characteristics as $T = T_1 + T_2 + T_3 + T_4 + T_5$.
Find the tolerances T_i, $i = 1, 2, 3, 4$, and 5, so that the total cost is minimized and the tolerance relationship is satisfied.

10. Consider the problem analyzed by Mansoor in his paper.[6] Please make the following change in the assumption made by him: X_i follows a beta distribution with parameters $\gamma = \eta = 2.0$, for all i. Keep all other assumptions. Derive all relevant equations and describe the steps of a solution procedure by which we can obtain the tolerances of the k components of an assembly.

11. Consider a dc circuit in which the voltage, V, across the points is required to be in the range 100 ± 2 volts. The specifications on the current I and the resistance R are $25 \pm T_I/2$ and $4 \pm T_R/2$, respectively. The voltage $V = IR$. The cost of manufacturing a resistance with tolerance T_R is $5/T_R$ and the cost of a current source with tolerance T_I is $2/T_I$. Determine the tolerances T_I and T_R so that the total cost is minimized.

12. An assembly consists of five components. The manufacturing costs of these components are given by $C_i = h_i T_i^{\alpha_i}$, where T_i is the tolerance on the characteristic of component i and α_i and h_i are given in the following table:

i	α_i	h_i
1	−1.00	1.00
2	−1.00	1.50
3	−1.00	1.80
4	−0.50	1.00
5	−1.00	1.00

The assembly tolerance should not exceed 0.02. The relationship among the assembly characteristic and the component characteristics is

$$X = \frac{X_1 \times X_2 \times X_3 \times X_4}{X_5}$$

The nominal sizes of X_1, X_2, X_3, X_4, and X_5 are 1.0, 1.5, 1.0, 2.0, and 1.0, respectively. Find the tolerances on X_i that minimize the total unit manufacturing cost and satisfy the tolerance relationship.

13. An assembly characteristic is related to the characteristics of the three components in the assembly by $X = 2X_1 - X_2 + X_3$. The processes available for manufacturing the components along with

the associated unit costs and the natural process tolerances are as follows:

Component 1			Component 2			Component 3		
Process	t	Unit Cost	Process	t	Unit Cost	Process	t	Unit Cost
1	0.001	$100	1	0.001	$120	1	0.002	$90
2	0.002	$80	2	0.003	$90	2	0.003	$80
3	0.003	$65	3	0.004	$80	3	0.005	$50

The tolerance on X is 0.006 and the ratio of the component tolerance to the natural process is 1.3 (G) for all components. Formulate (that is, define the decision variables and write the objective function and constraints) this problem as a zero–one integer programming problem to find T_i, $i = 1, 2$, and 3, that minimize the total unit cost of manufacture. (Do not solve for T_i).

14. An assembly characteristic is related to the characteristics of the three components in the assembly by $X = X_1 + X_2 + X_3$. The processes available for manufacturing the components along with the associated unit costs and the natural process tolerances are as follows:

Component 1		Component 2		Component 3	
Process	t	Process	t	Process	t
1	2.0	1	1.0	1	1.0
2	3.0	2	2.0	2	3.0
3	4.0	3	3.0	3	4.0

The tolerance on X is 7.0 and the ratio of the component tolerance to the natural process tolerance is 1.3 (G) for all components. The characteristic X_i follows a beta distribution with parameters $\gamma = \eta = 3.0$ for all i. You can assume that the distribution of X is approximately normal. Make all other assumptions made by Mansoor[6] and find (solve) the tolerances T_i for $I = 1, 2$, and 3, using his method so that the actual value of G is in the range (1.3–1.35). (This means that the actual G must not be less than 1.3 and not be greater than 1.35.) You need to find any one set of T_i that satisfy this requirement.

3

Loss Function

CONTENTS

3.1 Introduction

Let us consider the diameter of a shaft with specifications $1'' \pm 0.04''$, which means that the nominal value is $1''(B)$, the lower specification limit (LSL) is $0.96''$, and the upper specification limit (USL) is $1.04''$. Let the diameter (X) follow a density function $f(x)$ with a mean of μ and a variance of σ^2. Assume that the cost of reworking an oversized shaft is C_w and the cost of scrapping an undersized shaft is C_s. We will also assume that a shaft that is reworked will fall within the specification limits (this may not be realistic, but this assumption is being made to simplify the expression). Let the manufacturing cost per shaft be C_m. Then, the expected cost per shaft can

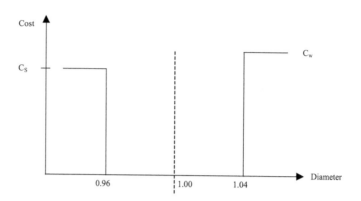

FIGURE 3.1
Cost versus diameter.

be written as:

$$E(TC) = C_m + C_s \int_{-\infty}^{LSL} f(x)\, dx + C_w \int_{USL}^{\infty} f(x)\, dx \tag{3.1}$$

In Eq. (3.1), C_m is a constant with respect to X (the diameter), so let us delete it from the expected cost expression. Figure 3.1 contains the plot of the cost per shaft as a function of X. This graph assumes that there is no cost (other than the manufacturing cost) incurred, as long as the diameter is between the LSL (0.96") and the USL (1.04"). It is not a valid approach, however, because all the shafts produced at 1.00" should carry a very low quality-liability cost that could even be $0 if they are perfect. They would mate with bearings or sleeves perfectly and not wear out prematurely. In other words, these shafts would not cause warranty costs, customer inconvenience, or loss of goodwill for the manufacturer.

Let us consider a shaft whose diameter is 0.970", which is still within the specification limits (0.96"–1.04") but on the loose side. This shaft will fit loosely with a bearing or sleeve (unless matched), increasing the probability of customer complaint and possibly failing prematurely. This example illustrates that even though the allowable range of diameters is 0.96"–1.04", the manufacturer must strive to keep the diameters as close as possible to 1", which becomes the target value.

Let us consider another example involving two types of resistors. The quality characteristic is the resistance that has a tolerance range from 950 to 1050 ohms. The nominal value is 1000 ohms. The histograms of the resistances of 50 resistors of each type are given in Figures 3.2a and b. It can be seen that even though the ranges of the resistances of both types are within the tolerance range, the width of the range of type-A resistors is much narrower than the width of the range of type-B resistors. It is obvious that customers prefer type-A resistors to type-B, because a randomly selected type-A resistor has a

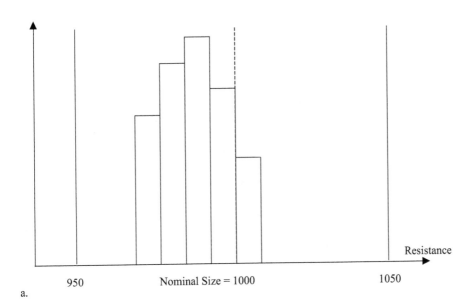

a.

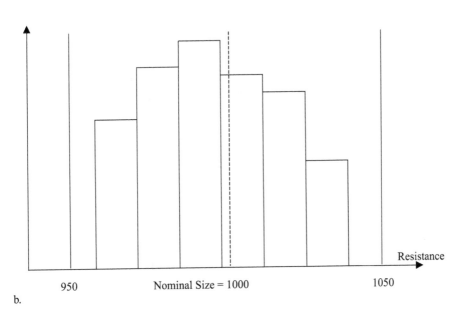

b.

FIGURE 3.2
(a) Histogram of resistances, type A. (b) Histogram of resistances, type B.

larger probability of being closer to the target value than a randomly selected type-B resistor.

Studies[10] have also shown that products with quality characteristics following normal distributions result in less failures, lower warranty costs, and higher customer satisfaction compared to products with quality characteristics following a uniform distribution, even though the range of this uniform distribution is within the tolerance range, resulting in zero proportion of defectives. This is because the uniform distribution has a larger variance than a truncated normal distribution with the same range.

The above examples emphasize the fact that the traditionally used quality metric—proportion of defectives that affect the internal failure costs—alone is not sufficient to measure the quality of a product. They also indicate that the parameters of the probability distribution of the quality characteristic of the product affect the performance of the product, which impacts the external failure costs. Even though external failure costs incurred after the product leaves the premises of the manufacturer are widely used in industries, there is no method available to predict these costs based on the parameters of the distribution of the quality characteristic. The loss function developed by Taguchi remedies this problem.[10]

3.2 Development of Loss Function

According to Taguchi, the cost of deviating from the target (1" in our first example) increases as the diameter moves away from the target. This cost is zero when the quality characteristic (X) is exactly equal to the target value, denoted by X_0 (see Figure 3.3). The cost of deviating from the target value is given by the loss function derived as follows.

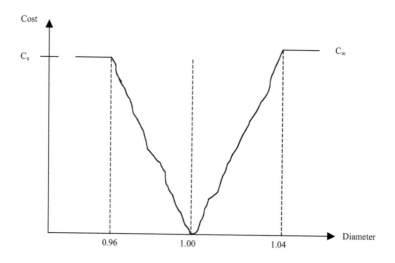

FIGURE 3.3
True cost curve.

The loss function, when the quality characteristic is X, is denoted by $L(X)$. This can be written as:

$$L(X) = L(X_0 + X - X_0) \tag{3.2}$$

Expanding the right-hand side using the Taylor series,

$$\left(f(X) = f(a + X - a) = f(a) + \frac{(X-a)}{1!} f'(a) + \frac{(X-a)^2}{2!} f''(a) + \cdots \right)$$

we obtain:

$$L(X) = L(X_0) + (X - X_0)L'(X_0) + \frac{(X-X_0)^2}{2!} L''(X_0) + \cdots, \tag{3.3}$$

where $L'(X_0)$ and $L''(X_0)$ are the first and second derivatives of $L(X)$, respectively, evaluated at X_0.

As per the assumption made earlier, the loss when the quality characteristic, X, is equal to its target value, X_0, is zero, and $L(X_0)$ is zero. Also, as the function $L(X)$ attains its minimum value when X is equal to X_0, its first derivative (slope) at X_0, $L'(X_0)$, is zero. Let us neglect the terms of third and higher orders. Then, Eq. (3.3) becomes:

$$L(X) = \frac{L''(X_0)}{2}(X - X_0)^2$$

$$= k'(X - X_0)^2, \quad LSL \leq X \leq USL \tag{3.4}$$

where $k' = \frac{L''(X_0)}{2}$ is a proportionality constant. According to Taguchi, this function represents the loss (in \$) incurred by the customer, the manufacturer, and society (due to warranty costs, customer dissatisfaction, etc.) caused by deviation of the quality characteristic from its target value. The dimension of $(X - X_0)^2$ is the square of the dimension of X. For example, if X is the diameter of a shaft measured in inches, then the dimension of $(X - X_0)^2$ is inch^2. As the dimension of $L(X)$ is \$, the dimension of the proportionality constant, k', has to be \$/(dimension of X)^2. The derivation of k' for various types of quality characteristics will be done later on.

Now we will derive the expected value of $L(X)$ given in Eq. (3.4):

$$E[L(X)] = E[k'(X - X_0)^2] = k'E[(X - X_0)^2], \text{ assuming the LSL and USL}$$

are contained within the range of X

$$= k'E[(X - \mu + \mu - X_0)^2]$$

$$= k'E[(X - \mu)^2 + 2(X - \mu)(\mu - X_0) + (\mu - X_0)^2]$$

$$= k'[E(X - \mu)^2 + 2(\mu - X_0)E(X - \mu) + E(\mu - X_0)^2] \tag{3.5}$$

where μ and X_0 are constants. As $\mathrm{Var}(X) = E(X - \mu)^2$, Eq. (3.5) can be written as:

$$
\begin{aligned}
E[L(X)] &= k'[\mathrm{Var}(X) + 2(\mu - X_0)[E(X) - \mu] + (\mu - X_0)^2] \\
&= k'[\sigma^2 + (\mu - X_0)^2]
\end{aligned}
\tag{3.6}
$$

where $E(X) = \mu$, and $\mathrm{Var}(X) = \sigma^2$. From Eq. (3.6):

$$
E[(X - X_0)^2] = [\sigma^2 + (\mu - X_0)^2]
\tag{3.7}
$$

We will derive the proportionality constant, k', and the estimates of $E[L(X)]$ for different types of quality characteristics in the next section.

3.3   Loss Functions for Different Types of Quality Characteristics

3.3.1   Nominal-the-Best Type (N Type)

3.3.1.1   *Equal Tolerances on Both Sides of the Nominal Size*

Tolerances for these types of characteristics are specified as $B \pm \Delta$, where B is the nominal value and hence is the target value, and Δ is the allowance on either side of the nominal size (the tolerance is 2Δ). The lower specification limit and the upper specification limit are $B - \Delta$ and $B + \Delta$, respectively. Let the rejection costs incurred by the manufacturer be C_s and C_w, when X is less than LSL and X is greater than USL, respectively.

When the costs C_s and C_w are equal, the loss function is

$$
\begin{aligned}
L(X) &= k'(X - X_0)^2, \quad \mathrm{LSL} \le X \le \mathrm{USL}, \\
&= 0, \text{ otherwise}
\end{aligned}
\tag{3.8}
$$

Let $C_s = C_w = C$. It is assumed that when $X = \mathrm{LSL}$ or when $X = \mathrm{USL}$, the loss, $L(X) = C$ may not be true. The resulting loss function is given in Figure 3.4:

$$
\begin{aligned}
L(\mathrm{LSL}) &= k'(\mathrm{LSL} - X_0)^2 = k'\Delta^2 \\
&= C; \text{ hence} \\
k' &= \frac{C}{\Delta^2}
\end{aligned}
\tag{3.9}
$$

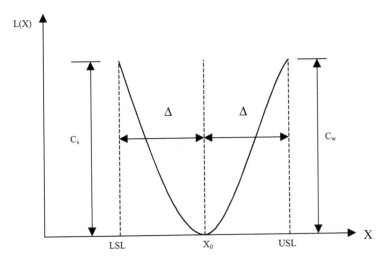

FIGURE 3.4
Loss when rejection costs are equal.

The expected value of the loss function defined in Eq. (3.8) is

$$E[L(X)] = k' \int_{LSL}^{USL} (x - X_0)^2 f(x)\, dx \tag{3.10}$$

$$= k'v^2 \tag{3.11}$$

where v^2 is called the *mean-squared deviation* and is equal to:

$$v^2 = \int_{LSL}^{USL} (x - X_0)^2 f(x)\, dx$$

$$= E(X - X_0)^2,\ \text{assuming that LSL and USL are within the range of } X$$

$$= [\sigma^2 + (\mu - X_0)^2],\ \text{per Eq. (3.7)} \tag{3.12}$$

The estimate of the expected loss defined in Eq. (3.11) is

$$E[L(\hat{X})] = k'\hat{v}^2 \tag{3.13}$$

where

$$\hat{v}^2 = \frac{1}{n} \sum_{i \in A} (X_i - X_0)^2 \tag{3.14}$$

The set A in Eq. (3.14) contains all observations in the range (LSL – USL). The right-hand side of Eq. (3.14) is the unbiased estimator of Eq. (3.12) which is also $[\sigma^2 + (\mu - X_0)^2]$. Also, n in Eq. (3.14) is the total number of observations in the sample batch collected to estimate the expected loss. Now we will show that the right-hand side of Eq. (3.14) is an unbiased estimator of $[\sigma^2 + (\mu - X_0)^2]$, assuming that all n observations are within the range (LSL – USL):

$$E\left[\frac{1}{n}\sum_{i=1}^{n}(X_i - X_0)^2\right] = \frac{1}{n}E\left[\sum_{i=1}^{n}(X_i^2 - 2X_iX_0 + X_0^2)\right]$$

$$= \frac{1}{n}\sum_{i=1}^{n}[E(X_i^2) - 2X_0E(X_i) + E(X_0^2)]$$

As $\sigma^2 = E(X_i^2) - [E(X_i)]^2$ and $E(X_i) = \mu$,

$$E\left[\frac{1}{n}\sum_{i=1}^{n}(X_i - X_0)^2\right] = \frac{1}{n}\sum_{i=1}^{n}[(\sigma^2 + \mu^2) - (2X_0\mu + X_0^2)]$$

$$= \frac{1}{n}[n\sigma^2 + n\mu^2 - 2X_0n\mu + nX_0^2]$$

$$= [\sigma^2 + \mu^2 - 2X_0\mu + X_0^2]$$

$$= [\sigma^2 + (\mu - X_0)^2] \tag{3.15}$$

While estimating $[\sigma^2 + (\mu - X_0)^2]$, it is natural to use $[S^2 + (\overline{X} - X_0)^2]$, in which S^2 is the sample variance (the unbiased estimator of σ^2) and $\overline{X}$ is the sample mean (the unbiased estimator of μ). But, it can be shown that $[S^2 + (\overline{X} - X_0)^2]$ is a biased estimator of $[\sigma^2 + (\mu - X_0)^2]$, as follows:

$$E[S^2 + (\overline{X} - X_0)^2] = E[S^2 + \overline{X}^2 - 2\overline{X}X_0 + X_0^2]$$

$$= E(S^2) + E(\overline{X}^2) - 2X_0E(\overline{X}) + X_0^2$$

As Var $(\overline{X}) = E(\overline{X}^2) - [E(\overline{X})]^2$ and $E(S^2) = \sigma^2$,

$$E[S^2 + (\overline{X} - X_0)^2] = \sigma^2 + \text{Var}(\overline{X}) + [E(\overline{X})]^2 - 2X_0E(\overline{X}) + X_0^2$$

As $E(\overline{X}) = \mu$ and Var $(\overline{X}) = \dfrac{\sigma^2}{n}$,

$$E[S^2 + (\overline{X} - X_0)^2] = \sigma^2 + \frac{\sigma^2}{n} + \mu^2 - 2X_0\mu + X_0^2$$

$$= \sigma^2 + (\mu - X_0)^2 + \frac{\sigma^2}{n} \tag{3.16}$$

It can be seen from Eq. (3.16) that the estimator $[S^2 + (\bar{X} - X_0)^2]$ overestimates $[\sigma^2 + (\mu - X_0)^2]$ by $\frac{\sigma^2}{n}$. This bias will decrease as n increases.

Example 3.1

The specification limits for the resistance of a resistor are 1000 ± 50 ohms. A resistor with resistance outside these limits will be discarded at a cost of $0.50. A sample of 15 resistors yielded the following observations (in ohms).

1020 1040 980 1000 980 1000 1010 1000 1030 970 1000 960 990 1040 960

Estimate the expected loss per resistor.

$$LSL = 950$$
$$USL = 1050$$
$$B = X_0 = 1000$$
$$\Delta = 50$$
$$C_s = C_w = \$0.50$$

$$L(X) = k'(X-1000)^2, \quad 950 \le X \le 1050$$
$$= 0, \quad \text{otherwise}$$

From Eq. (3.9):

$$k' = \frac{0.50}{50^2} = 2 \times 10^{-4}$$

From Eq. (3.10):

$$E[L(X)] = k' \int_{950}^{1050} (x - 1000)^2 f(x)\, dx$$

The estimate of $E[L(X)] = k'\hat{\sigma}^2$, where, as per Eq. (3.14):

$$\hat{\sigma}^2 = \frac{1}{15} \sum_{i \in A} (X_i - 1000)^2$$

The set A contains all the observations in the range $(950 - 1050)$. Hence,

$$
\begin{aligned}
\hat{v}^2 &= \frac{1}{15}[(1020 - 1000)^2 + (1040 - 1000)^2 + (980 - 1000)^2 + (1000 - 1000)^2 \\
&\quad + (980 - 1000)^2 + (1000 - 1000)^2 + (1010 - 1000)^2 + (1000 - 1000)^2 \\
&\quad + (1030 - 1000)^2 + (970 - 1000)^2 + (1000 - 1000)^2 + (960 - 1000)^2 \\
&\quad + (990 - 1000)^2 + (1040 - 1000)^2 + (960 - 1000)^2] \\
&= \frac{9600}{15} = \frac{0.96 \times 10^4}{15}
\end{aligned}
$$

Now,

$$
E[L(\hat{X})] = 2 \times 10^{-4} \times \frac{0.96 \times 10^4}{15} = \$\,0.13
$$

When the costs C_s and C_w are not equal, the loss function is

$$
\begin{aligned}
L(X) &= k_1' (X - X_0)^2, \quad \text{LSL} \leq X \leq X_0 \\
&= k_2' (X - X_0)^2, \quad X_0 \leq X \leq \text{USL} \\
&= 0, \quad \text{otherwise}
\end{aligned}
\tag{3.17}
$$

It is assumed that when $X = \text{LSL}$, the loss $L(X) = C_s$, and when $X = \text{USL}$, the associated loss $L(X) = C_w$. The resulting loss function is given in Figure 3.5.

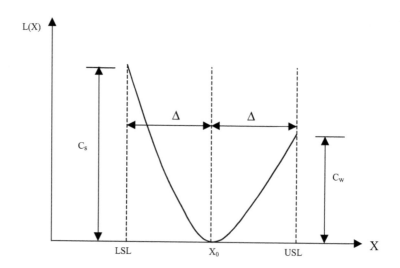

FIGURE 3.5
Loss when costs are unequal.

Based upon these assumptions:

$$L(\text{LSL}) = k_1'(\text{LSL} - X_0)^2 = k_1' \Delta^2$$
$$= C_s$$

hence:

$$k_1' = \frac{C_s}{\Delta^2} \tag{3.18}$$

Similarly,

$$L(\text{USL}) = k_2'(\text{USL} - X_0)^2 = k_2' \Delta^2$$
$$= C_w$$

hence:

$$k_2' = \frac{C_w}{\Delta^2} \tag{3.19}$$

Now the expected value of the loss function defined in Eq. (3.17) is

$$E[L(X)] = k_1' \int_{\text{LSL}}^{x_0} (x - X_0)^2 f(x)\, dx + k_2' \int_{x_0}^{\text{USL}} (x - X_0)^2 f(x)\, dx \tag{3.20}$$

$$= k_1' v_1^2 + k_2' v_2^2 \tag{3.21}$$

where:

$$v_1^2 = \int_{\text{LSL}}^{x_0} (x - X_0)^2 f(x)\, dx \tag{3.22}$$

is the part of the mean-squared deviation in the range from LSL to X_0, and

$$v_2^2 = \int_{X_0}^{\text{USL}} (x - X_0)^2 f(x)\, dx \tag{3.23}$$

is the part of mean-squared deviation in the range from X_0 to USL. The estimate of the expected loss defined in Eq. (3.21) is

$$E[L(\hat{X})] = k_1' \hat{v}_1^2 + k_2' \hat{v}_2^2 \tag{3.24}$$

where:

$$\hat{v}_1^2 = \frac{1}{n} \sum_{i \in A_1} (X_i - X_0)^2 \tag{3.25}$$

which estimates Eq. (3.22), and

$$\hat{v}_2^2 = \frac{1}{n} \sum_{i = A_2} (X_i - X_0)^2 \tag{3.26}$$

which estimates Eq. (3.23). In Eqs. (3.25) and (3.26), n is the total number of observations in the sample batch collected to estimate the expected loss. In Eq. (3.25), the set A_1 contains all observations in the range (LSL − X_0). In Eq. (3.26), the set A_2 contains all observations in the range (X_0− USL).

Example 3.2

A manufacturer of a component requires that the tolerance on the outside diameter be 5 ± 0.006". Defective components that are oversized can be reworked at a cost of $5.00 per piece. Undersized components are scrapped at a cost of $10.00 per piece. The following outside diameters were obtained from a sample batch of 20 components:

5.003,   5.000,   4.999,   5.000,   5.003,   5.002,   5.001,   4.998,   5.006,

5.004,   4.998,   5.001,   5.000,   4.996,   4.995,   4.998,   5.004,   5.000

5.006,   5.002

Estimate the expected loss per piece.

$\qquad$ LSL = 4.994

$\qquad$ USL = 5.006

$\qquad$ $B = X_0 = 5.000$

$\qquad$ $C_s = \$10.00$

$\qquad$ $C_w = \$5.00$

$$L(X) = k_1'(X - X_0)^2, \quad 4.994 \le X \le 5.000$$
$$= k_2'(X - X_0)^2, \quad 5.000 \le X \le 5.006$$
$$= 0, \quad \text{otherwise}$$

From Eq. (3.18),

$$k_1' = \frac{10}{0.006^2} = \frac{10 \times 10^6}{36}$$

Loss Function

From Eq. (3.19),

$$k_2' = \frac{5}{0.006^2} = \frac{5 \times 10^6}{36}$$

The expected value of the loss function per Eqs. (3.20) and (3.21) is

$$E[L(X)] = k_1' \int_{4.994}^{5} (x-5)^2 f(x)\, dx + k_2' \int_{5}^{5.006} (x-5)^2 f(x)\, dx$$

$$= k_1' v_1^2 + k_2' v_2^2$$

The estimate of the expected loss per Eq. (3.24) is

$$E[L(\hat{X})] = k_1' \hat{v}_1^2 + k_2' \hat{v}_2^2$$

where, per Eq. (3.25),

$$\hat{v}_1^2 = \frac{1}{n} \sum_{m \in A_1} (X_i - 5.000)^2$$

$$= \frac{1}{20} \begin{bmatrix} (5.000 - 5.000)^2 + (4.999 - 5.000)^2 + (5.000 - 5.000)^2 \\ + (4.998 - 5.000)^2 + (4.998 - 5.000)^2 + (5.000 - 5.000)^2 \\ + (4.996 - 5.000)^2 + (4.995 - 5.000)^2 + (4.998 - 5.000)^2 \end{bmatrix}$$

$$= \frac{1}{20} 54 \times 10^{-6} = 2.7 \times 10^{-6}$$

As per Eq. (3.26):

$$\hat{v}_2^2 = \frac{1}{20} \sum_{i \in A_2} (X_i - 5.000)^2$$

$$= \frac{1}{20} \begin{bmatrix} (5.003 - 5.000)^2 + (5.003 - 5.000)^2 + (5.002 - 5.000)^2 \\ + (5.001 - 5.000)^2 + (5.006 - 5.000)^2 + (5.004 - 5.000)^2 \\ + (5.001 - 5.000)^2 + (5.004 - 5.000)^2 + (5.006 - 5.000)^2 \\ + (5.002 - 5.000)^2 \end{bmatrix}$$

$$= \frac{1}{20} 132 \times 10^{-6} = 6.6 \times 10^{-6}$$

So,

$$E[L(\hat{X})] = \frac{10}{36} \times 10^6 \times 2.7 \times 10^{-6} + \frac{5}{36} \times 10^6 \times 6.6 \times 10^{-6}$$

$$= \$1.67 \text{ per piece}$$

If the manufacturer produces 50,000 units per month, then the expected loss per month is $50,000 \times 1.67 = \$83,500.00$.

3.3.1.2 Unequal Tolerances on Both Sides of the Nominal Size

Tolerances are specified as $B_{-\Delta_1}^{+\Delta_2}$. Hence, the LSL is $B - \Delta_1$, the USL is $B + \Delta_2$, and the nominal size (which is the target value), is B:

$$
\begin{aligned}
L(X) &= k_1'(X - X_0)^2, \quad X_0 - \Delta_1 \leq X \leq X_0 \\
&= k_2'(X - X_0)^2, \quad X_0 \leq X \leq X_0 + \Delta_2 \\
&= 0, \quad \text{otherwise}
\end{aligned}
\tag{3.27}
$$

As before, it is assumed that when $X = \text{LSL}$, the loss $L(X) = C_s$, and when $X = \text{USL}$, the associated loss $L(X) = C_w$. The resulting loss function is given in Figure 3.6. Based upon these assumptions,

$$
\begin{aligned}
L(\text{LSL}) &= k_1'(\text{LSL} - X_0)^2 = k_1'\Delta_1^2 \\
&= C_s
\end{aligned}
$$

hence,

$$
k_1' = \frac{C_s}{\Delta_1^2}
\tag{3.28}
$$

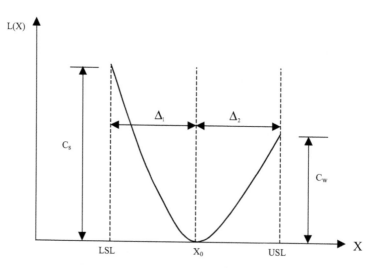

FIGURE 3.6
Loss when tolerances are unequal.

Similarly,

$$L(USL) = k_2'(USL - X_0)^2 = k_2'\Delta_2^2$$
$$= C_w$$

hence,

$$k_2' = \frac{C_w}{\Delta_2^2} \tag{3.29}$$

The expected value of the loss function and its estimate are the same as given in Eqs. (3.20) through (3.26), with $LSL = B - \Delta_1$ and $USL = B + \Delta_2$.

Example 3.3

The specifications for the thickness of a gauge block are $1''{}^{+0.002}_{-0.001}$. Defective blocks that are undersized have to be scrapped at a cost of \$12.00 a piece, and the blocks that are oversized can be reworked at a cost of \$5.00 a piece. The following are the actual thickness values of 15 blocks:

1.001 0.999 0.999 1.002 1.000 1.001 1.002 0.999 1.001 0.999
1.001 1.000 0.999 1.002 0.999

Estimate the expected loss per piece.

$LSL = 0.999''$
$USL = 1.002''$
$B = X_0 = 1.000''$
$C_s = \$12.00$
$C_w = \$5.00$

$$L(X) = k_1'(X - X_0)^2, \quad 0.999'' \le X \le 1.000''$$
$$= k_2'(X - X_0)^2, \quad 1.000'' \le X \le 1.002''$$
$$= 0, \quad \text{otherwise}$$

From Eq. (3.28),

$$k_1' = \frac{8}{0.001^2} = 8 \times 10^6$$

and from Eq. (3.29),

$$k_2' = \frac{5}{0.002^2} = \frac{5 \times 10^6}{4}$$

The expected value of the loss function per Eqs. (3.20) and (3.21) is

$$E[L(X)] = k_1' \int_{0.999}^{1.000} (x-1)^2 f(x)\,dx + k_2' \int_{1.000}^{1.002} (x-1)^2 f(x)\,dx$$

$$= k_1 v_1^2 + k_2 v_2^2$$

The estimate of the expected loss per Eq. (3.24) is

$$E[L(\hat{X})] = k_1'\,\hat{v}_1^2 + k_2'\,\hat{v}_2^2$$

where per Eq. (3.25):

$$\hat{v}_1^2 = \frac{1}{n} \sum_{m \in A_1} (X_i - 1.000)^2$$

$$= \frac{1}{15} \begin{bmatrix} (0.999 - 1.000)^2 + (0.999 - 1.000)^2 + (1.000 - 1.000)^2 \\ + (0.999 - 1.000)^2 + (0.999 - 1.000)^2 + (1.000 - 1.000)^2 \\ + (0.999 - 1.000)^2 + (0.999 - 1.000)^2 \end{bmatrix}$$

$$= \frac{6}{15} \times 10^{-6}$$

Per Eq. (3.26):

$$\hat{v}_2^2 = \frac{1}{15} \sum_{i \in A_2} (X_i - 1.000)^2$$

$$= \frac{1}{15} \begin{bmatrix} (1.001 - 1.000)^2 + (1.002 - 1.000)^2 + (1.001 - 1.000)^2 \\ + (1.002 - 1.000)^2 + (1.001 - 1.000)^2 + (1.001 - 1.000)^2 \\ + (1.002 - 1.000)^2 \end{bmatrix}$$

$$= \frac{16}{15} \times 10^{-6}$$

So,

$$E[L(\hat{X})] = 8 \times 10^6 \times \frac{6}{15} \times 10^{-6} + \frac{5}{4} \times 10^6 \times \frac{16}{15} \times 10^{-6}$$

$$= \$4.53 \text{ per piece}$$

3.3.2 Smaller-the-Better Type (S Type)

Tolerances for this type of characteristics are specified as $X \le \Delta$, where the upper specification limit is Δ. There is no lower specification limit for these characteristics. It is assumed that the quality characteristic X is non-negative. Let the rejection costs incurred by the manufacturer be C_w when X is greater

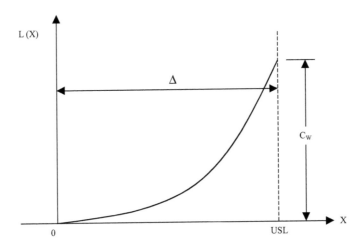

FIGURE 3.7
Loss for smaller-the-better type (S type).

than USL. Some examples of this type of characteristic include impurity, shrinkage, noise level, flatness, surface roughness, roundness, and wear. The implied target value, X_0, is 0; hence, the loss function is

$$L(X) = k'X^2, \quad X \le \text{USL}$$
$$= 0, \quad \text{otherwise} \tag{3.30}$$

As in the case of other quality characteristics, it is assumed that the loss when $X = \text{USL}$ is C_w. The resulting loss function is given in Figure 3.7.
  As per the assumption,

$$L(\text{USL}) = k'\Delta^2 = C_w$$

hence,

$$k' = \frac{C_w}{\Delta^2}$$

The expected value of the loss function defined in Eq. (3.30) is

$$E[L(X)] = k'\int_0^{\text{USL}} x^2 f(x)\,dx \tag{3.31}$$

$$= k'v^2 \tag{3.32}$$

where v^2 is the mean-squared deviation and is equal to:

$$v^2 = \int_0^{USL} x^2 f(x)\, dx \tag{3.33}$$

The estimate of the expected loss defined in Eqs. (3.31) and (3.32) is

$$E[L(\hat{X})] = k'\hat{v}^2 \tag{3.34}$$

where

$$\hat{v}^2 = \frac{1}{n}\sum_{i\in A} X_i^2 \tag{3.35}$$

where n is the sample size and A is the set containing all observations in the interval $(0-\Delta)$.

Example 3.4
A manufacturer of ground shafts requires that the surface roughness of the surface of each shaft be within 10 units. The loss caused by out-of-tolerance conditions is \$20.00 per piece. The surface roughness data on 10 shafts are given below:

$$10\ \ 5\ \ 6\ \ 2\ \ 4\ \ 8\ \ 1\ \ 3\ \ 5\ \ 1$$

Compute the expected loss per shaft.

$$L(X) = k'X^2, \quad X \le 10$$
$$= 0, \quad \text{otherwise}$$

As $C_w = \$20.00$ and $\Delta = 10$,

$$k' = \frac{20}{10^2} = 0.20$$

$$\hat{v}^2 = \frac{1}{10}[10^2 + 5^2 + 6^2 + 2^2 + 4^2 + 8^2 + 1^2 + 3^2 + 5^2 + 1^2] = 19.1$$

$$E[L(X)] = 0.20 \times 19.1$$
$$= \$3.82$$

3.3.3 Larger-the-Better Type (L Type)

Tolerances for this type of characteristics are specified as $X \ge \Delta$, where the lower specification limit is Δ. There is no upper specification limit for these characteristics. Let the rejection costs incurred by the manufacturer be C_w, when X is less than the LSL. Some of the examples of this type of characteristic are tensile strength, compressive strength, and miles per gallon. The implied target value, X_0, is ∞ and the loss function $L(X) = k(X - X_0)^2$ is equal to ∞ for all values of X. To eliminate this problem, the L-type characteristic is transformed to an S-type characteristic using the transformation $Y = 1/X$. Now, Y becomes an S-type characteristic with an upper specification limit $= 1/\Delta$, hence the loss function for Y is

$$L(Y) = k'Y^2, \quad Y \leq \frac{1}{\Delta}$$

$$= 0, \quad \text{otherwise}$$

(3.36)

As in the case of other quality characteristics, it is assumed that the loss when $X = \Delta$ or when $Y = 1/\Delta$ is C_w. In Eq. (3.36),

$$k' = \frac{C_w}{\left(\frac{1}{\Delta}\right)^2} = C_w \Delta^2$$

(3.37)

Now let us transform the variable Y back to the original variable X using the transformation, $X = 1/Y$. Then, from Eq. (3.36), the loss function of X is

$$L(X) = \frac{k'}{X^2}, \quad X \geq \Delta$$

$$= 0, \quad \text{otherwise}$$

(3.38)

where k' is as per Eq. (3.37). The loss function is given in Figure 3.8. The expected value of the loss function defined in Eq. (3.38) is

$$E[L(X)] = k' \int_{LSL}^{\infty} \frac{1}{x^2} f(x)\, dx$$

(3.39)

$$= k'v^2$$

(3.40)

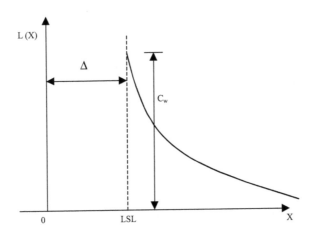

FIGURE 3.8
Loss for larger-the-better type (L type).

where v^2 is the mean-squared deviation and is equal to:

$$v^2 = \int_{LSL}^{\infty} \frac{1}{x^2} f(x)\, dx \tag{3.41}$$

The estimate of the expected loss defined in Eqs. (3.39) and (3.40) is

$$E[L(\hat{X})] = k'\hat{v}^2 \tag{3.42}$$

where

$$\hat{v}^2 = \frac{1}{n} \sum_{i\in A} \frac{1}{X_i^2} \tag{3.43}$$

where n is the sample size and A is the set containing all observations $\geq \Delta$, which is the LSL.

Example 3.5

The producer of a certain steel beam used in construction requires that the strength of the beam be more than 30,000 lb/in.^{2} The cost of a defective beam is $600.00. The annual production rate is 10,000 beams. The following data (lb/in.^{2}) were obtained from destructive tests performed on 10 beams:

$$\begin{array}{ccccc}
40{,}000 & 41{,}000 & 60{,}000 & 45{,}000 & 65{,}000 \\
35{,}000 & 41{,}000 & 51{,}000 & 60{,}000 & 49{,}000
\end{array}$$

What is the expected loss per year?

$$L(X) = \frac{k'}{X^2}, \quad X \geq 30{,}000$$
$$= 0, \quad \text{otherwise}$$

$C_w = \$600.00$

$\Delta = 30{,}000$

$k' = 600 \times (30{,}000)^2 = 5400 \times 10^8$

$$\hat{v}^2 = \frac{10^{-8}}{10}\left[\frac{1}{40^2} + \frac{1}{41^2} + \frac{1}{60^2} + \frac{1}{45^2} + \frac{1}{65^2} + \frac{1}{35^2} + \frac{1}{41^2} + \frac{1}{51^2} + \frac{1}{60^2} + \frac{1}{49^2}\right]$$

$$= 0.004718 \times 10^{-8}$$

$$E[L(\hat{X})] = 5400 \times 10^8 \times 0.004718 \times 10^{-8} = \$214.20 \text{ per unit}$$

The loss per year, then, is $214.20 \times 10{,}000 = \$2{,}142{,}000.00$.

3.4 Robust Design using Loss Function

3.4.1 Methodology

The expected value of loss function as per (3.6) is

$$E[L(x)] = k'[\sigma^2 + (\mu - x_0)^2]$$

Let X be the quality characteristic of the assembly with a mean μ, variance σ^2, and target value X_0. In order to minimize the expected loss of X, we should:

1. Make the mean of X, $\mu = X_0$.
2. Minimize the variance of X, σ^2.

Let us assume that the quality characteristic of the assembly X is a known function of the characteristics of the components of the assembly, $X_1, X_2, X_3, \ldots, X_k$, assuming k components in the assembly, given by:

$$X = e(X_1, X_2, \ldots, X_k) \tag{3.44}$$

As X is a function of the component characteristics $X_1, X_2, \ldots, X_k$, it is clear that the mean and variance of X (μ and σ^2) can be controlled by controlling (selecting, setting, etc.) the means of $X_1, X_2, \ldots, X_k$ denoted by $\mu_1, \mu_2, \ldots, \mu_k$ and the variances $\sigma_1^2, \sigma_2^2, \ldots, \sigma_k^2$. As we saw in earlier chapters, the variances (σ_i^2) depend upon the processes, and the means (μ_i) depend upon the process setting. The robust design that we are going to discuss now determines the values of $\mu_1, \ldots, \mu_k$ for given values of $\sigma_1^2, \sigma_2^2, \ldots, \sigma_k^2$, so that $E(X) = \mu = X_0$ and $Var(X) = \sigma^2$ is minimized. The means can be set equal to the respective nominal sizes, $B_1, B_2, \ldots, B_k$.

The equations derived are valid for any probability density function of X_i^5. Let us expand $e(x_1, \ldots, x_k)$ about $\mu_1, \mu_2, \ldots, \mu_k$ using the Taylor series and neglect terms of order three and higher:

$$X = e(X_1, X_2, \ldots, X_k)$$

$$= e(\underline{\mu}) + \sum_{i=1}^{k} g_i(\underline{\mu})(X_i - \mu_i) + \frac{1}{2}\sum_{i=1}^{k}\sum_{j=1}^{k} h_{ij}(\underline{\mu})(X_i - \mu_i)(X_j - \mu_j) \tag{3.45}$$

where μ is the row vector equal to $\underline{\mu} = [\mu_1, \mu_2, \ldots, \mu_k]$ and $g_i(\underline{\mu}) = \partial^2 e(\cdot)/\partial X_i$ evaluated at $\underline{\mu} = [\mu_1, \mu_2, \ldots, \mu_k]$. Also, $h_{ij}(\underline{\mu}) = \partial^2 e(\cdot)/\partial X_i \partial X_j$ evaluated at $\underline{\mu} = [\mu_1, \mu_2, \ldots, \mu_k]$.

Taking the expected value of both sides of Eq. (3.45):

$$\mu = E(X)$$
$$= E[e(X_1, X_2, \ldots, X_k)]$$
$$= E[e(\underline{\mu})] + E\left[\sum_{i=1}^{k} g_i(\underline{\mu})(X_i - \mu_i)\right] + \frac{1}{2}E\left[\sum_{i=1}^{k}\sum_{j=1}^{k} h_{ij}(\underline{\mu})(X_i - \mu_i)(X_j - \mu_j)\right]$$
$$= e(\underline{\mu}) + \sum_{i=1}^{k} g_i(\underline{\mu})[E(X_i) - E(\mu_i)] + \frac{1}{2}\sum_{i=j}^{k}\sum_{i=j}^{k} h_{ij}(\underline{\mu})E[(X_i - \mu_i)(X_j - \mu_j)]$$

$$(3.46)$$

as $g_i(\underline{\mu})$, $h_{ij}(\underline{\mu})$, and $e(\underline{\mu})$ are constants.
$E(X_i) = \mu_i$, so $E(X_i) - \mu = 0$. Also, when $i = j$,

$$E[(X_i - \mu_i)(X_j - \mu_j)] = E[(X_i - \mu_i)^2]$$
$$= \text{Var}(X_i) = \sigma_i^2$$

and when $i \neq j$,

$$E[(X_i - \mu_i)(X_j - \mu_j)] = \text{covariance of } (X_i, X_j)$$

Combining these results:

$$\sigma_{ij} = \text{Cov}(X_i, X_j), \quad \text{if } i \neq j$$
$$= \sigma_i^2, \quad \text{if } i = j$$

The covariance of X_i and X_j is 0, if X_i and X_j are independent.
    Now, from Eq. (3.46), the mean of X is

$$\mu = e(\underline{\mu}) + \frac{1}{2}\sum_{i=1}^{k}\sum_{j=1}^{k} h_{ij}(\underline{\mu})\sigma_{ij} \qquad (3.47)$$

The quantity $(\mu - X_0)$ is called the bias and is to be minimized. From Eq. (3.47), it is

$$(\mu - X_0) = e(\underline{\mu}) + \frac{1}{2}\sum_{i=1}^{k}\sum_{j=1}^{k} h_{ij}(\underline{\mu})\sigma_{ij} - X_0 \qquad (3.48)$$

One of our objectives is to make the right-hand side of Eq. (3.48) equal to 0, which yields:

$$e(\underline{\mu}) + \frac{1}{2}\sum_{i=1}^{k}\sum_{j=1}^{k} h_{ij}(\underline{\mu})\sigma_{ij} = X_0 \qquad (3.49)$$

Now let us derive an expression for the variance of X, σ^2:

$$\sigma^2 = E[(X - \mu)^2]$$

From Eqs. (3.45) and (3.47),

$$X - \mu = e(\underline{\mu}) + \sum_{i=1}^{k} g_i(\underline{\mu})(X_i - \mu_i) + \frac{1}{2}\sum_{i=1}^{k}\sum_{j=1}^{k} h_{ij}(\underline{\mu})(X_i - \mu_i)(X_j - \mu_j)$$

$$- e(\underline{\mu}) - \frac{1}{2}\sum_{i=1}^{k}\sum_{i=j}^{k} h_{ij}(\underline{\mu})\sigma_{ij}$$

$$= \sum_{i=1}^{k} g_i(\underline{\mu})(X_i - \mu_i) + \frac{1}{2}\sum_{i=1}^{k}\sum_{j=1}^{k} h_{ij}[(X_i - \mu_i)(X_j - \mu_j) - \sigma_{ij}] \qquad (3.50)$$

$(X - \mu)^2$ is approximated by $[\sum_{i=1}^{k} g_i(\underline{\mu})(X_i - \mu_i)]^2$.[5] Hence, the variance of X is

$$\sigma^2 = E\left[\sum_{i=1}^{k} g_i(\underline{\mu})(X_i - \mu_i)\right]^2 \qquad (3.51)$$

As:

$$\left[\sum_{i=1}^{k} a_i b_i\right]^2 = \sum_{i=1}^{k} a_i b_i \left[\sum_{j=1}^{k} a_j b_j\right] = \sum_{i=1}^{k}\sum_{j=1}^{k} a_i b_i a_j b_j = \sum_{i=1}^{k}\sum_{j=1}^{k} a_i a_j b_i b_j$$

$$\left[\sum_{i=1}^{k} g_i(\underline{\mu})(X_i - \mu_i)\right]^2 = \sum_{i=1}^{k}\sum_{j=1}^{k} g_i(\underline{\mu})g_j(\underline{\mu})(X_i - \mu_i)(X_j - \mu_j)$$

Hence, Eq. (3.51) is

$$\sigma^2 = E\left[\sum_{i=1}^{k}\sum_{j=1}^{k} g_i(\underline{\mu})g_j(\underline{\mu})(X_i - \mu_i)(X_j - \mu_j)\right]$$

$$= \sum_{i=1}^{k}\sum_{j=1}^{k} g_i(\underline{\mu})g_j(\underline{\mu})E[(X_i - \mu_i)(X_j - \mu_j)]$$

$$= \sum_{i=1}^{k}\sum_{j=1}^{k} g_i(\underline{\mu})g_j(\underline{\mu})\sigma_{ij}] \qquad (3.52)$$

where:

$$\sigma_{ij} = \text{Cov}(X_i, X_j), \quad \text{if } i \neq j$$
$$= \sigma_i^2, \quad \text{if } i = j$$

Now the problem of robust design can be formulated as follows.

Find the means $\mu_1, \mu_2, \ldots, \mu_k$ that can be set equal to the nominal sizes $B_1, \ldots, B_k$ so as to minimize:

$$\sigma^2 = \sum_{i=1}^{k}\sum_{j=1}^{k} g_i(\underline{\mu})g_j(\underline{\mu})\sigma_{ij} \quad \text{(given in Eq. (3.52))}$$

subject to:

$$e(\underline{\mu}) + \frac{1}{2}\sum_{i=1}^{k}\sum_{j=1}^{k} h_{ij}(\underline{\mu})\sigma_{ij} = X_0 \quad \text{(given in Eq. (3.49))} \qquad (3.53)$$

Other applicable constraints can be added to the formulation in Eq. (3.53). It is assumed that in Eq. (3.53), the variances of X_i, σ_i^2, and $\text{Cov}(X_i, X_j)$ are known. These can be estimated using the data collected on the X_i. For larger-the-better type (L-type) characteristics, the target value is infinity, hence the constraint given by Eq. (3.49) can be changed to a maximization objective function. Similarly, for smaller-the-better (S-type) characteristics, the target value is zero, hence the constraint may be changed to a minimization objective function.

The formulation given in Eq. (3.53) is a nonlinear programming problem with both nonlinear objective function and constraint. The following example illustrates the solution of this formulation for a simple assembly.

Example 3.6

The assembly characteristic of a product, X, is related to the component characteristics, X_1 and X_2, by the following relation: $X = X_1 X_2$. The target value of X is 35.00. Formulate the robust design problem to find the means (nominal sizes) of X_1 and X_2. Assume that X_1 and X_2 are independent and that the variances of X_1 and X_2, which are σ_1^2 and σ_2^2, are known. It is given that $e(X_1, X_2) = X_1 X_2$.

First we find $g_i(\underline{\mu})$ and $h_{ij}(\underline{\mu})$ for all i and j:

$$g_1(\underline{\mu}):\frac{\delta e(\cdot)}{\delta X_1} = X_2; \quad g_1(\underline{\mu}) = \mu_2$$

$$g_2(\underline{\mu}):\frac{\delta e(\cdot)}{\delta X_2} = X_1; \quad g_2(\underline{\mu}) = \mu_1$$

$$h_{11}(\underline{\mu}):\frac{\delta e(\cdot)}{\delta X_1} = X_2; \quad \frac{\delta e(\cdot)}{\delta X_1 \delta X_1} = 0; \quad h_{11}(\underline{\mu}) = 0$$

$$h_{21}(\underline{\mu}):\frac{\delta e(\cdot)}{\delta X_1} = X_2; \quad \frac{\delta^2 e(\cdot)}{\delta X_2 \delta X_1} = 1; \quad h_{21}(\underline{\mu}) = 1 = h_{12}(\underline{\mu})$$

$$h_{22}(\underline{\mu}):\frac{\delta e(\cdot)}{\delta X_2} = X_1; \quad \frac{\delta^2 e(\cdot)}{\delta X_2^2} = 0; \quad h_{22}(\underline{\mu}) = 0$$

As X_1 and X_2 are independent, $\text{Cov}(X_1, X_2) = \sigma_{12} = \sigma_{21} = 0$.

FORMULATION

Find μ_1 and μ_2 so as to minimize:

$$\sigma^2 = \sum_{i=1}^{2}\sum_{j=1}^{2} g_i(\underline{\mu})g_j(\underline{\mu})\sigma_{ij}$$

$$= g_1(\underline{\mu})g_1(\underline{\mu})\sigma_1^2 + g_1(\underline{\mu})g_2(\underline{\mu})\sigma_{12} + g_2(\underline{\mu})\sigma_{21} + g_2(\underline{\mu})g_2(\underline{\mu})\sigma_2^2$$

$$= \mu_2^2\sigma_1^2 + \mu_1^2\sigma_2^2$$

subject to:

$$e(\underline{\mu}) + \frac{1}{2}\sum_{i=1}^{2}\sum_{j=1}^{2} h_{ij}(\underline{\mu})\sigma_{ij} = 35$$

$$e(\mu_1,\mu_2) + \frac{1}{2}[h_{11}(\underline{\mu})\sigma_1^2 + h_{12}(\underline{\mu})\sigma_{12} + h_{21}(\underline{\mu})\sigma_{21} + h_{22}(\underline{\mu})\sigma_2^2] = 35$$

$$\mu_1\mu_2 = 35$$

That is, find μ_1 and μ_2 so as to minimize $\mu_2^2\sigma_1^2 + \mu_1^2\sigma_2^2$ subject to $\mu_1\mu_2 = 35$.

This simpler problem can be solved using the LaGrange multiplier method,[1] which converts the above constrained problem into an unconstrained problem. The unconstrained problem here is to find μ_1 and μ_2 so as to minimize $F = \mu_2^2\sigma_1^2 + \mu_1^2\sigma_2^2 + \lambda(\mu_1\mu_2 - 35)$, where λ is the LaGrange multiplier. The three unknowns—μ_1, μ_2, and λ—can be found by setting the partial derivatives of F with respect to these variables set equal to 0:

$$\frac{\partial F}{\partial\mu_1} = 2\mu_1\sigma_2^2 + \lambda\mu_2 = 0; \quad \lambda = -\frac{2\mu_1\sigma_2^2}{\mu_2} \qquad\text{(i)}$$

$$\frac{\partial F}{\partial\mu_2} = 2\mu_2\sigma_1^2 + \lambda\mu_1 = 0; \quad \lambda = -\frac{2\mu_2\sigma_1^2}{\mu_1} \qquad\text{(ii)}$$

$$\frac{\partial F}{\partial\lambda} = \mu_1\mu_2 - 35 = 0; \quad \mu_1\mu_2 = 35. \qquad\text{(iii)}$$

From Eqs. (i) and (ii), $\mu_1/\mu_2 = \sigma_1/\sigma_2$ and using Eq. (iii):

$$\frac{\mu_1}{\left(\frac{35}{\mu_1}\right)} = \frac{\sigma_1}{\sigma_2}$$

which gives

$$\mu_1 = \sqrt{35\left(\frac{\sigma_1}{\sigma_2}\right)} \quad \text{and} \quad \mu_2 = \sqrt{35\left(\frac{\sigma_2}{\sigma_1}\right)}$$

If $\sigma_1 = 1.0$ and $\sigma_2 = 10.0$, then $\mu_1 = \sqrt{35 \times 0.1} = 1.87$ and $\mu_2 = \sqrt{35 \times 10} = 18.71$. The optimum variance (optimum value of the objective function) is $(18.71)^2 \times 1.0 + (1.87)^2 \times 100 = 699.75$, and the optimum standard deviation is 26.45. Nonlinear programming software such as GINO can be used for solving difficult formulations.

The usual practice is to arbitrarily set the values of the nominal sizes (means of the component characteristics) that yield the nominal size for the assembly characteristic, ignoring the variances of the component characteristics. When these nominal sizes are arbitrarily set, then the resulting variance of the assembly characteristic according to Eq. (3.52) can be reduced only by reducing the variances of the component characteristics, which could be very time consuming and expensive. For example, let us assume that the design engineer arbitrarily sets the values of μ_1 and μ_2 as 5 and 7, respectively, which yield the mean of $X = 35$, the target value, but the resulting variance of X is

$$\sigma^2 = \mu_2^2\sigma_1^2 + \mu_1^2\sigma_2^2 = 7^2 \times 1 + 5^2 \times 100 = 2549$$

and the standard deviation is 50.49, which is larger than the optimum value of 26.45. The only way to reduce this large variance is to reduce σ_1^2 and σ_2^2 by sorting and matching the components or tightening the tolerances of the components (assuming the processes are fixed), which are both expensive to the manufacturer. Out of all the infinite combinations of μ_1 and μ_2 that yield the target value of X (35 in this example), robust design formulation selects the optimum values of μ_1 and μ_2 that minimize the variance of X. This minimizes the expected loss, wherein lies the advantage of robust design, which meets the requirement of concurrent design.

3.4.2 Some Recent Developments in Robust Design

As was evident from the discussion in the preceding section, robust design improves the quality of a product by adjusting the means of the component characteristics so that the variance of the assembly characteristic is minimized and the mean of the assembly characteristic is equal to its target value. The reduction in variance is equivalent to decreasing the sensitivity of the assembly characteristic to the noise or uncontrollable factors in the design, manufacturing, and functional stages of the product. Robust design, a quality assurance methodology, is applied to the design stage, where the nominal sizes of the components are determined. Since Taguchi's initial work in this area, many researchers have expanded his contribution. Oh[6] provides a very good overview of this work. Efforts in the design stages of products have made a dramatic impact on the quality of these products. The design phase is divided into three parts: system design, parameter design, and tolerance design.[11]

3.4.2.1   System Design

In this step, the basic prototype of the product is developed to perform the required functions of the final product, and the materials, parts, and manufacturing and assembly systems are selected.

3.4.2.2   Parameter Design

The optimum means of the design parameters of the components are selected in this phase so that the product characteristic is insensitive to the effect of noise factors. Robust design plays a major role in this step.

The key to developing the formulation of the robust design problem in Eq. (3.53) is the function, $e(X_1, X_2, ..., X_k)$, in Eq. (3.44) which relates the assembly characteristic X to the component characteristics, $X_1, X_2, ..., X_k$. In most real-life problems, that function may not be available. Taguchi recommends design of experiments in such cases.[7] He developed signal-to-noise ratios (known as S/N ratios) to combine the objective function of minimizing the variance and making the mean equal to the target value. This is more helpful for larger-the-better (L-type) characteristics with a target value of infinity and smaller-the-better type (S-type) characteristics with a target value of 0. Unal and Dean,[11] Chen et al.,[2] and Scibilia et al.[8] are some of the authors who have extended the work of Taguchi. Multiple criteria optimization has been considered by Chen et al.[2] and Song et al.[9] Genetic algorithms combined with the finite element method have been used in robust design by Wang et al.,[12] and evolutionary algorithms have been used by Wiesmann et al.[13]

3.4.2.3   Tolerance Design

This step is carried out only if the variation of the product characteristic achieved in parameter design is not satisfactory. Here, optimum tolerances that minimize the total cost are determined. The optimization techniques used in this step include response surface methodology, integer programming, nonlinear programming, and simulation.[3,4]

3.5   References

1. Bazara, M.S., Sherali, H.D., and Shetty, C.M., *Nonlinear Programming: Theory and Algorithms*, 2nd ed., John Wiley & Sons, New York, 1993.
2. Chen, M., Chiang, P., and Lin, L., Device robust design using multiple-response optimization technique, 5th International Conference on Statistical Metrology, 46–49, 2000.
3. Feng, Chang-Xue and Kusiak, A., Robust tolerance design with the integer programming approach, *J. Manuf. Sci. Eng.*, 119, 603–610, 1997.

4. Jeang, A. and Lue, E., Robust tolerance design by response surface methodology, *Int. J. Prod. Res.*, 15, 399–403, 1999.
5. Oh, H.L., Modeling Variation to Enhance Quality in Manufacturing, paper presented at the Conference on Uncertainty in Engineering Design, Gaithersburg, MD, 1988.
6. Oh, H.L., A changing paradigm in quality, *IEEE Trans. Reliability*, 44, 265–270, 1995.
7. Phadke, M.S., *Quality Engineering Using Robust Design*, Prentice Hall, New York, 1989.
8. Scibilia, B., Kobi, A., Chassagnon, R., and Barreau, A., An application of the quality engineering approach reconsidered, *IEEE Trans.*, 943–947, 1999.
9. Song, A., Pattipati, K.R., and Mathur, A., Multiple criteria optimization for the robust design of quantitative parameters, *IEEE Trans.*, 2572–2577, 1994.
10. Taguchi, G., Elsayed, E.A., and Hsiang, T., *Quality Engineering in Production Systems*, McGraw-Hill, New York, 1989.
11. Unal, R. and Dean, E.B., *Design for Cost and Quality: The Robust Design Approach*, National Aeronautics & Space Administration, Washington, D.C., 1995.
12. Wang, H.T., Liu, Z.J., Low, T.S., Ge, S.S., and Bi, C., A genetic algorithm combined with finite element method for robust design of actuators, *IEEE Trans. Magnetics*, 36, 1128–1131, 2000.
13. Wiesmann, D., Hammel, U., and Back, T., Robust design of multilayer optical coatings by means of evolutionary algorithms, *IEEE Trans. Evolutionary Comp.*, 2, 162–167, 1998.

3.6 Problems

1. The specification limits of the internal diameter of a sleeve are 1.5 ± 0.005. The costs of scrapping an oversized sleeve and of reworking an undersized sleeve are $50.00 and $20.00, respectively. The internal diameters of 20 sleeves randomly selected from the production line are as follows:

1.502	1.499	1.500	1.498	1.497	1.504	1.503
1.503	1.500	1.499	1.498	1.501	1.497	1.496
1.497	1.504	1.501	1.500	1.496	1.499	

Estimate the expected loss per sleeve.

2. The thickness of a spacer has to lie between 0.504 and 0.514. The cost of rejecting a sleeve is $10.00. Ten spacers were measured, yielding the following readings:

0.508	0.509	0.510	0.512	0.506
0.513	0.510	0.508	0.511	0.512

Estimate the expected loss per month, if the production quantity per month is 100,000.

3. The tensile strength of a component has to be greater than or equal to 20,000 tons per square inch(ton/in.^{2}). The cost of failure of a component with strength less than 20,000 ton/in.^{2} is $300.00. Tests of 10 components yielded the following tensile strengths:

21,000	30,000	35,000	40,000	25,000
32,000	28,000	45,000	50,000	30,000

Estimate the expected loss per component.

4. The surface roughness of a surface plate cannot exceed 10 units. The cost of rework of a plate with surface roughness greater than 10 units is $100.00. Twelve surface plates had the following surface roughness values:

2   4   5   7   6   3   1   9   7   8

Estimate the expected loss per surface plate.

5. Suppose that you asked an operator to collect n observations and estimate the mean-squared deviation, V^2, of a nominal-the-best type characteristic. The operator, by mistake, computed the sample variance, S^2, instead. By the time the operator realized his mistake, he lost all the values of the individual observations (that is, the X_i). He could give you only the following values:

$S^2 = 10.0$

$\overline{X} = 1.5$ (sample mean)

$X_0 = 1.4$ (target value)

$n = 10$ (sample size)

He also reported to you that all 10 observations were within the range (LSL – USL).

Compute an estimate of the mean-squared deviation using the above information.

6. An assembly consisting of three components has the following relationship between its assembly characteristic and the component characteristics:

$$X = \frac{E X_1^4}{(X_1 + X_2)^3 X_3}$$

where E is a known constant. Assume that the standard deviation of X_i is σ_i for $i = 1, 2,$ and 3 and is given. Formulate this as a robust design problem in which the decision variables are the nominal sizes of the X_i (B_i), which are equal to the respective means (μ_i). Assume that the X_i are independent of each other.

4

Process Capability

CONTENTS

4.1 Introduction

From earlier discussions in previous chapters, it can be seen that almost all the quality control problems can be solved if the following conditions for manufacturing the product are met:

1. The quality characteristics are within the appropriate specification/ tolerance limits determined based on customers' requirements.

2. The variability of the quality characteristics is minimized as much as possible.

3. The mean of each quality characteristic is as close as possible to the target value of the characteristic.

Condition 1 eliminates the products that are defectives (outside the specification limits), while conditions 2 and 3 reduce the proportion of defectives and enhance the consistency of the performance of the products around the target values of their quality characteristics. Taguchi's loss function captures the effects of both conditions 2 and 3.

Process capability analysis relates the mean and the variance of the distribution of a quality characteristic to the specification/tolerance limits and gives numerical measures of the extent to which the above conditions are met. In this chapter, some of the commonly used indexes to measure process capability are discussed.

4.2   Preliminaries

Process capability matches the capability of the process with the range of the specification/tolerance interval for a given quality characteristic. The capability of the process is measured in terms of the range of all possible values of the quality characteristic. This is the range of all the values of the characteristic of the component/product in a given lot or batch if the user is interested in measuring the capability of the process with respect to that lot. In this case, the population consists of the values of the characteristic in the lot only so the size of the population is finite. On the other hand, it can be the range of all possible values of the characteristic that the process can generate under some specified conditions, if the user wants to measure the capability of the process under those conditions. Here, the population is the set of all possible values that the process can generate under those conditions, thus its size is infinite.

It is obvious that if the range of the distribution of the quality characteristic is less than the range of the specification (tolerance) interval, then the number of nonconforming products or defectives will be zero. On the other hand, if the range of the characteristic is wider than the specification range, then the process is bound to produce defectives. The exact range of the values of a characteristic of the items in a lot (finite population) or of the range of all possible values that a process can generate (infinite population) can be obtained exactly only if all these values are measured (ignoring measurement error). This requires 100% inspection, which may not be feasible in all cases or, even if feasible, may not be economical. The only alternative is to take a sample batch of parts from the lot or from the process while it is running in the in-control state (Chapter 8), measure the values of the characteristic of all the items in that batch, and make inference about the range of all the values in

TABLE 4.1

Areas for Different Ranges Under Standard Normal Curve

Range	Covered Within the Range (%)	Outside the Range (%)	Outside the Range (ppm)
$(\mu - 1\,\sigma)$ to $(\mu + 1\,\sigma)$	68.26	31.74	317,400
$(\mu - 2\,\sigma)$ to $(\mu + 2\,\sigma)$	95.44	4.56	45,600
$(\mu - 3\,\sigma)$ to $(\mu + 3\,\sigma)$	99.73	0.27	2700
$(\mu - 4\,\sigma)$ to $(\mu + 4\,\sigma)$	99.99366	0.00634	63.4
$(\mu - 5\,\sigma)$ to $(\mu + 5\,\sigma)$	99.9999426	0.0000574	0.574
$(\mu - 6\,\sigma)$ to $(\mu + 6\,\sigma)$	99.9999998	0.0000002	0.002

the population (finite or infinite) using the sample information. This is possible because the ranges of the values of the probability distributions can be expressed as functions of their parameters, especially their standard deviations. As an example, the ranges pertaining to a normal distribution, the most commonly used distribution in statistical quality control, are given in the Table 4.1. The results in this table are valid only if:

1. The probability distribution of the quality characteristic generated by the process is exactly normal.
2. The exact values of the mean of the distribution, μ, and the standard deviation of the distribution, σ, are known.

For example, consider a turning lathe, which is used to machine the outside diameter of a batch of shafts. Let the distribution of the outer diameters generated by the lathe be normally distributed with a mean $\mu = 1''$ and a standard deviation $\sigma = 0.0005''$. Then, 99.9937% of the shafts processed on this machine will have their outside diameters within the interval from $0.998''$ $(1 - [4 \times 0.0005])$ to $1.002''$ $(1 + [4 \times 0.0005])$, and 0.0063% of the outside diameters generated by the machine will be outside this interval. The width of this range is $8 \times 0.0005 = 0.004''$.

As the theoretical limits of a normal distribution are $-\infty$ and $+\infty$, which are not realizable in real-life situations, it does not have a finite interval containing 100% of the values of the quality characteristic. It has been the convention to take the interval from $(\mu - 3\sigma)$ to $(\mu + 3\sigma)$, the width of which is equal to 6σ, as the benchmark against which to measure the process capability. This is known as the *6-sigma spread*, which indicates the width of this interval. It can be seen from the table that this interval contains only 99.73% of all possible values of the characteristic and that 0.27% of the values (in other words, 2700 defective parts out of one million, or 2700 parts per million) fall outside this interval. The recent trend among many industries is to consider the interval from $(\mu - 6\sigma)$ to $(\mu + 6\sigma)$ as the measure of the capability of the process. This interval, the width of which is 12σ, contains 99.9999998% of the values of the quality characteristic generated by the process. The percentage of values outside this interval is 0.0000002, which translates to 0.002 per million (0.002 parts per million, or ppm).

Even if the distribution of the characteristic is normal, the results in the table are valid only if the exact values of the mean of the distribution μ and the standard deviation σ are used to compute the limits of the intervals. Exact values of these parameters can be known only if all the values in the population (which could be a batch or a lot or all possible values that can be generated by a process under a set of specified conditions) are known. In the absence of such 100% inspection, which may not be feasible and/or economical, the parameters are estimated from the results of 100% inspection of a sample batch, which contains a small subset of the values from the population. The sample mean $\overline{X}$ is used to estimate the mean of the distribution μ (known as the population mean), and either R/d_2 (where R is the sample range and d_2 is a constant) or s/c_4 (where s is the sample standard deviation and c_4 is a constant) is used to estimate the standard deviation of the distribution σ (known also as the population standard deviation). The constants d_2 and c_4 are given for various values of n in Table A.4 in the Appendix. Out of these, $\overline{X}$ is an unbiased estimator of μ with the minimum variance, and even though both R/d_2 and s/c_4 are unbiased estimators of σ, the variance of s/c_4 is less than that of R/d_2. But, the desirable properties of these estimators—namely, unbiasedness and minimum variance—cannot minimize the error in estimation unless the sample batch from which these statistics are computed is truly random. A sample batch is said to be random, if every item in the original population has the same chance of being included in the sample batch. In simple terms, this implies that the sample batch must truly represent the population it is taken from, especially its central tendency and variability.

Selection of random samples for the purpose of estimating the capability of a running process becomes complicated because of the possibility of changes in the parameters of the population of the quality characteristic in the time interval during which the parts are manufactured. Here, the issues of sample size and length of the time interval required to collect the observations in the sample batch are very important. Let us consider the following example.

Example 4.1

It takes 10 seconds for an N.C. lathe to machine the outside diameter of a pin. Sample batches, each consisting of five pins, were collected every 10 minutes. The diameters of six such batches are given in Table 4.2. As the time interval between successive sample batches was 10 minutes, these batches

TABLE 4.2

Diameters of Pins in Example 4.1

Sample batch #1	1.000	0.999	0.999	1.000	1.000
Sample batch #2	1.001	1.000	1.000	0.999	1.001
Sample batch #3	1.001	1.001	1.002	1.000	1.000
Sample batch #4	1.002	1.001	1.001	1.002	1.001
Sample batch #5	1.002	1.003	1.002	1.003	1.002
Sample batch #6	1.003	1.002	1.001	1.002	1.002

TABLE 4.3

Sample Statistics in Example 4.1

Time	Batch #	s	s/c_4	$\bar{x}$
0.0	1	0.000535	0.000569	0.9996
10.0	2	0.000817	0.000869	1.0002
20.0	3	0.000817	0.000869	1.0008
30.0	4	0.000618	0.000657	1.0014
40.0	5	0.000618	0.000657	1.0024
50.0	6	0.000756	0.000804	1.0020
Grand average			0.000738	1.001

TABLE 4.4

Two Sets of Estimates in Example 4.1

Parameter	Estimate 1	Estimate 2
μ	1.001	1.001
σ	0.000738	0.001329

were collected over a period of 50 minutes. The sample statistics of these six batches are given in Table 4.3.

Let us consider two sets of estimates for the mean μ and the standard deviation σ of the process. One set of estimates is obtained from the grand averages of the s/c_4 and $\bar{x}$ of the six sample batches given in Table 4.3 and the other set consists of the s/c_4 and $\bar{x}$ values calculated by considering all 30 observations as one single sample batch. That is, the second set of estimates is computed as follows:

$$\bar{x} = \frac{(1.000 + 0.999 + 0.999 + \cdots + 1.001 + 1.002 + 1.002)}{30}$$

$$= 1.001 \text{ (same as the grand average in Table 4.3)}$$

$$s = \sqrt{\frac{(1.000 - 1.001)^2 + \cdots + (1.002 - 1.001)^2}{(30 - 1)}}$$

$$= 0.001329$$

As the value of c_4 for a sample size of 30 is very close to 1.000, this s value need not be divided by c_4 in order to get an unbiased estimate of σ. These two sets of estimates are given in Table 4.4.

As the estimate for μ is the same in the two sets, let us consider the estimates for σ. The second estimate is almost twice as much as the first estimate. The question is, which of these estimates truly represents the variability present in the diameters of the 300 pins? It is obvious that the larger value (0.001329) is the true estimate of the standard deviation of the 300 diameters, because it includes the long-term variability present in the diameters, whereas the first estimate captures only the short-term variability present in the sample batch of size 5, which is the natural variability inherent in the process. It can be seen from Table 4.3 that the mean of the population increases during the 50 minutes, and the estimate 0.001329 captures this increase in addition to representing the natural variability of the process present within a batch of size 5. Process control techniques (discussed in Chapter 8) can be used to detect and prevent such increases in the mean. Only when the process control techniques confirm that the process is in control should the parameters for computing process capability indexes be estimated.

4.3 Process Capability Indexes and Their Limitations

In this section, we will study some of the capability indexes used in industries.

4.3.1 Process Capability

Process capability is simply the range that contains all possible values of a specified quality characteristic generated by a process under a given set of conditions. As we discussed earlier, in the case of a normal distribution, there is no finite range that contains 100% of the values, hence, the range containing 99.73% of the values is taken as the benchmark, which is equal to $6\times$ the standard deviation. That is,

$$\text{Process Capability} = 6\sigma \qquad (4.1)$$

The recent trend is to take the process capability equal to either $8 \times$ the standard deviation that contains 99.9937% or $12 \times$ the standard deviation that contains 99.9999998% of the values. This number can be used in the initial selection of a machine or a process. It must be noted that the process capability of a process depends upon the levels of the process parameters (for example, cutting speed, feed, etc. in the case of a CNC lathe), thus the same process can have more than one process capability value.

The limitation of this index is due to the assumption that the related quality characteristic is normally distributed. The deviation of the shape of the distribution from normality will affect the range containing a certain percentage of the values.

4.3.2 Process Capability Ratio

The process capability ratio (PCR, or C_p) compares the tolerance specified on the characteristic with the process capability defined in Section 4.3.1 and gives an indication of the proportion or percentage of rejection. For nominal-the-best type of characteristics,

$$\text{PCR or } C_p = \frac{(\text{USL} - \text{LSL})}{6\sigma} \qquad (4.2)$$

where USL and LSL are the upper and lower specification limits of the quality characteristic, respectively. The quantity in the denominator, 6σ, signifies the fact that 99.73% of the values generated by the process is contained within

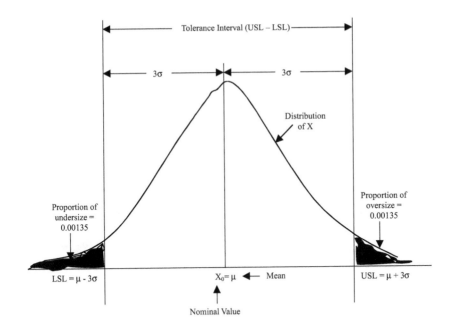

FIGURE 4.1
Proportion of defectives when $C_p = 1.0$.

the interval $(\mu - 3\sigma, \mu + 3\sigma)$, assuming normal distribution for the quality characteristic generated by the process. This implies that if the capability of the process is such that 99.73% of the values of a quality characteristic falls within the limits of the tolerance interval, then its PCR is equal to 1. In other words, the percentage of rejection is equal to 0.27%, or the number of defectives out of 1 million parts is 2700 (2700 ppm), if the PCR is 1 and the mean of the distribution of the characteristic is equal to the mid-point of the tolerance interval, which is the nominal size. This is represented in Figure 4.1. Similarly, if 99.9999998% of the values falls within the tolerance interval, then the PCR value is equal to 2.0, because (USL − LSL) = 12σ, assuming normal distribution. In this case, the percentage rejection is 0.0000002 (0.002 defectives per 1 million parts, or 0.002 ppm).

It is simple to develop a formula to estimate the percentage of defectives generated by a process from its PCR value, assuming that the probability distribution of the quality characteristic is normal. Consider Figure 4.2, in which the upper and lower limits of the two-sided tolerance interval are marked on the distribution of the characteristic, denoted by X. The total proportion of defectives is the sum of the shaded areas under the distribution curve to the left of the lower specification limit and to the right of the upper specification limit. Probabilistically, this is equal to:

$$p = \text{Prob}[X < \text{LSL}] + \text{Prob}[X > \text{USL}]$$

$$= P\left[Z < \frac{\text{LSL} - \mu}{\sigma}\right] + P\left[Z > \frac{\text{USL} - \mu}{\sigma}\right] \qquad (4.3)$$

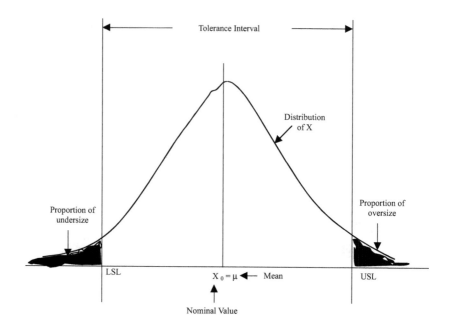

FIGURE 4.2
Proportion of defectives.

If the mean of the distribution coincides with the mid-point of the tolerance interval (which is the nominal size) as per the assumption, then the two shaded areas are equal and the proportion of defectives, p, can be written as:

$$p = 2 \times P\left[Z < \frac{\text{LSL} - \mu}{\sigma}\right] \quad \text{or} \quad 2 \times P\left[Z > \frac{\text{USL} - \mu}{\sigma}\right] \quad (4.4)$$

As the mean μ is equal to $(\text{LSL} + \text{USL})/2$, each term on the right-hand side of the above equation can be rewritten as follows:

$$2 \times P\left[Z < \frac{\text{LSL} - \mu}{\sigma}\right] = 2 \times P\left[Z < \frac{\text{LSL} - \frac{(\text{LSL} + \text{USL})}{2}}{\sigma}\right]$$

$$= 2 \times P\left[Z < \frac{\text{LSL} - \text{USL}}{2\sigma}\right]$$

$$= 2 \times P\left[Z < \frac{-6 \times \sigma \times C_p}{2\sigma}\right], \quad \text{as} \quad C_p = \frac{(\text{USL} - \text{LSL})}{6\sigma}$$

per Eq. (4.2)

$$= 2 \times P[Z < -3C_p] \quad (4.5)$$

Similarly,

$$2 \times \text{Prob}\left[Z > \frac{\text{USL} - \mu}{\sigma}\right] = 2 \times \text{Prob}\left[Z > \frac{\text{USL} - \frac{(\text{LSL} + \text{USL})}{2}}{\sigma}\right]$$

$$= 2 \times \text{Prob}\left[Z > \frac{(\text{USL} - \text{LSL})}{2\sigma}\right]$$

$$= 2 \times \text{Prob}[Z > 3C_p] \tag{4.6}$$

Now, combining Eqs. (4.4), (4.5), and (4.6), p can be written as:

$$p = 2 \times P[Z < -3C_p] \quad \text{or} \quad 2 \times P[Z > 3C_p] \tag{4.7}$$

Example 4.3

The specification limits for the inside diameter of a hole are (0.995", 1.005"). The standard deviation of the inside diameters generated by a lathe machine selected to process this component is estimated to be 0.002". Compute the C_p index of this process and estimate the proportion of defectives; assume that the inside diameters generated by the machine follow normal distribution and that the mean of the distribution is equal to the nominal size, which is 1.000":

$$C_p = \frac{(\text{USL} - \text{LSL})}{6 \times \sigma}$$

$$= \frac{(1.005 - 0.995)}{6 \times 0.002}$$

$$= 0.8333$$

The estimate of the proportion of defectives is

$$p = 2 \times P[Z > 3C_p]$$
$$= 2 \times P[Z > 3 \times 0.8333]$$
$$= 2 \times P[Z > 2.4999]$$
$$= 2 \times 0.0062 \text{ (from standard normal tables)}$$
$$= 0.0122$$

The PCR index for quality characteristics with just one limit—either upper or lower limit—is computed a little differently than the index for quality characteristics with both upper and lower limits, even though the index conveys the same information in both cases. In the case of quality characteristics with just one limit, the tolerance interval of interest is the distance from the limit to the mean of the distribution, μ. This distance is equal to $(\text{USL} - \mu)$ in the case of characteristics with just the upper limit only (smaller-the-better) and $(\mu - \text{LSL})$ in the case of characteristics with just the lower limit only (larger-the-better). These distances are compared with one half of the spread

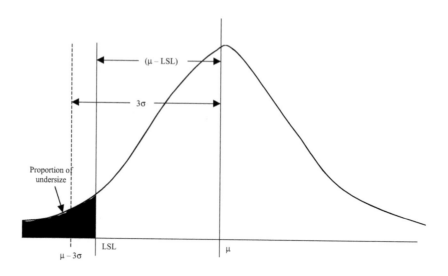

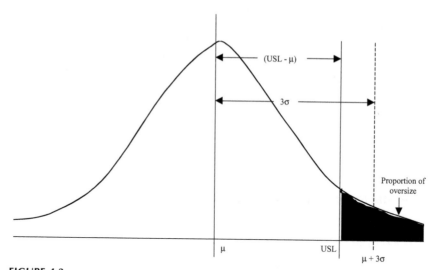

FIGURE 4.3
(a) C_p index for larger-the-better characteristics; (b) C_p index for smaller-the-better characteristics.

of the characteristics, which is equal to 3σ, as shown in Figures 4.3a and b. The PCR index is, then,

$$C_p = \begin{cases} \dfrac{(USL - \mu)}{3\sigma}, & \text{for smaller-the-better type characteristics} \\ \dfrac{(\mu - LSL)}{3\sigma}, & \text{for larger-the-better type characteristics} \end{cases} \quad (4.8)$$

It can be seen by comparing Eqs. (4.2) and (4.8) that the major difference between the C_p values for characteristics with either one or two specification limits is the presence of the mean, μ, in the expression of C_p for characteristics with one limit. This eliminates the necessity of making any assumption about the location of the mean, while estimating the proportion of defectives from the C_p value for the characteristics with just one specification limit. The proportion of defectives, p, can be estimated as:

$$p = P[X < LSL] \quad \text{or} \quad P[X > USL]$$

$$= P\left[Z < \frac{LSL - \mu}{\sigma}\right] \quad \text{or} \quad P\left[Z > \frac{USL - \mu}{\sigma}\right]$$

$$= P\left[Z < \frac{-3\sigma C_p}{\sigma}\right] \quad \text{or} \quad P\left[Z > \frac{3\sigma C_p}{\sigma}\right]$$

$$\text{as } C_p = \frac{\mu - LSL}{3\sigma} \quad \text{or} \quad \frac{USL - \mu}{3\sigma}, \text{ according to Eq. (4.8)}$$

$$= P[Z < -3C_p] \quad \text{or} \quad P[Z > 3C_p] \tag{4.9}$$

The only assumption required in order for the above equation to be valid is the normality of the distribution of the quality characteristic.

Example 4.4

The roughness of the ground surface of a component cannot exceed 0.02 units. A random sample of components ground by a surface-grinding machine yielded the following estimates:

Mean roughness = 0.01.

Standard deviation = 0.003.

Compute the C_p index of this process and estimate the proportion of defectives expected to be generated by the process, assuming that the surface roughness measurements follow normal distribution.

This is an example of a characteristic with only the upper specification limit (USL). From Eqs. (4.8) and (4.9),

$$C_p = \frac{USL - \mu}{3\sigma}$$

$$p = P[Z > 3C_p]$$

Using the estimates of μ and σ in the above equations, we get:

$$C_p = \frac{0.02 - 0.01}{3 \times 0.003}$$

$$= 1.111$$

$$p = P[Z > 3 \times 1.111]$$

$$= P[Z > 3.333]$$

$$= 0.0004$$

The proportion of defectives estimated above translates to 0.04% defectives, or 400 defective parts per 1 million ground parts (400 ppm).

Example 4.5

The tensile strength of a welded joint used in construction has to be at least equal to 40 tons per square inch (ton/in.^{2}). A random sample of 100 joints welded by a welding machine yielded the following estimates of the parameters of the distribution of the tensile strengths:

Mean = 65 ton/in.^{2}.

Standard deviation = 8.2 ton/in.^{2}.

Compute the C_p index and estimate the proportion of defectives generated by the machine, assuming that the distribution of the tensile strengths is normal.

This is an example of a characteristic with only the lower specification limit (larger-the-better type of characteristic). From Eqs. (4.8) and (4.9),

$$C_p = \frac{\mu - LSL}{3\sigma}$$

$$p = P[Z < -3C_p]$$

Replacing μ and σ in the above relations by their estimates results in:

$$C_p = \frac{65 - 40}{3 \times 8.2}$$

$$= 1.012$$

$$p = P[Z < -3 \times 1.012]$$

$$= P[Z < -3.036]$$

$$= 0.0012$$

The proportion of defectives computed above can also be expressed as 0.12%, or 1200 defectives per 1 million (1200 ppm), welded joints produced on the machine.

The limitations of C_p in the case of characteristics with both lower and upper specification limits are due to the assumptions made while estimating the proportion of defectives from the index: (1) the distribution of the characteristic is normal, and (2) the mean is equal to the nominal size. The limitation of C_p in the case of characteristics with just one lower or upper limit is only due to the assumption that the distribution of the characteristic is normal. The estimates of the proportion of defectives as per Eqs. (4.7) and (4.9) are not valid if the distribution of the characteristic is not normal.

Even if the distribution of the characteristic is normal, the C_p index will not present a true picture regarding the proportion of defectives if the mean is not equal to the nominal size in the case of characteristics with both lower and upper limits. This is clearly explained in Figure 4.4, which contains the

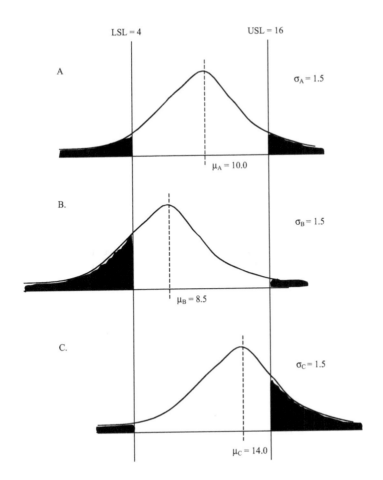

FIGURE 4.4
C_p indexes for processes with different means.

distributions of three processes—A, B, and C— which manufacture the same
component with a quality characteristic having lower and upper specification limits equal to 4 and 16, respectively. The nominal size is equal to 10.

All three processes have the same standard deviation, which is equal to 1.5 units,
thus the process capability ratio value for all these processes is equal to:

$$C_p = \frac{(16-4)}{6 \times 1.5} = 1.33$$

The estimate of the proportion of defectives using Eq. (4.7) is

$$p = 2 \times \text{Prob}[Z > 3 \times 1.33]$$
$$= 2 \times \text{Prob}[Z > 4.00]$$
$$= 2 \times 0.0000317 \quad \text{(from Table A.5 in the Appendix)}$$
$$= 0.0000634 \ (63.4 \text{ ppm})$$

The true proportion of defectives for these processes, however, are as follows:

Process A:

$$\text{Mean } (\mu) = 10(\text{nominal size})$$
$$p = P[X < 4] + P[X > 16]$$
$$= P\left[Z < \frac{4-10}{1.5}\right] + P\left[Z > \frac{16-10}{1.5}\right]$$
$$= P[Z < -4]P[Z > 4]$$
$$= 2 \times 0.0000317 \quad (\text{from Table A.5 in the Appendix})$$
$$= 0.0000634 \ (63.4 \text{ ppm})$$

This is the same as the estimate obtained earlier using Eq. (4.7), because the assumption that the mean is equal to the nominal size is satisfied in process A.

Process B:

$$\text{Mean}(\mu) = 8.5$$
$$p = P[X < 4] + P[X > 16]$$
$$= P\left[Z < \frac{4-8.5}{1.5}\right] + P\left[Z > \frac{16-8.5}{1.5}\right]$$
$$= P[Z < -3]P[Z > 5]$$
$$= 0.001350 + 0.000000287$$
$$= 0.00135029 \ (1350.287 \text{ ppm})$$

Process C:

$$\text{Mean}(\mu) = 14$$
$$p = P[X < 4] + P[X > 16]$$
$$= P\left[Z < \frac{4-14}{1.5}\right] + P\left[Z > \frac{16-14}{1.5}\right]$$
$$= P[Z < -6.67]P[Z > 1.33]$$
$$= 0 + 0.0918$$
$$= 0.0918 \ (91,800 \text{ ppm})$$

As can be seen from this example, the deviation of the mean from the nominal size greatly affects the proportion of the number of defectives produced by the process. This effect is not captured by the C_p index because of the assumption that the mean is equal to the nominal size when estimating the proportion of defectives from the index value. The next index to be discussed was developed to remedy this problem in the C_p index.

4.3.3 C_{pk} Index

The C_{pk} index was introduced to take care of the limitation of the C_p index, which assumes that the mean of the distribution is equal to the nominal size while estimating the proportion of defectives in the case of nominal-the-best type of characteristics. The reasoning behind the expression used for computing the C_{pk} index can best be explained through the example used to illustrate the limitation of the C_p index in the previous section (Figure 4.4). The proportion of defectives for nominal-the-best type of characteristics depends upon the distances of the mean of the distribution from both the limits. Hence, for such a characteristic, an ideal index that indicates process capability and can be used to estimate the proportion of defectives should contain some information about both these distances. But, if one is constrained to select only one of these distances, then the value to choose would be the minimum of the two distances, because the minimum represents the worst case in terms of the proportion of defectives. For process C in Figure 4.4, this would be the distance between the upper limit and the mean (USL $- \mu$), because for process C the mean is closer to the upper limit than the lower limit. For process B, this distance would be ($\mu -$ LSL), as the mean is closer to the lower limit. This distance should be divided by some multiple of the process standard deviation, σ, as in the case of the C_p index. The logical divisor is 3σ, because in this index only part of the range USL $-$ LSL is considered, as opposed to the entire range of USL $-$ LSL, as in the case of the C_p index in which the divisor is 6σ. Now, the expression for the C_{pk} index can be written as:

$$C_{pk} = \frac{\text{Minimum}[(\mu - \text{LSL}), (\text{USL} - \mu)]}{3\sigma} \qquad (4.10)$$

The C_{pk} index for characteristics with only one specification limit (smaller-the-better and larger-the-better types of characteristics) is the same as the C_p index given in Eq. (4.8):

$$C_{pk} = \begin{cases} \dfrac{(\text{USL} - \mu)}{3\sigma} & \text{for smaller-the-better type} \\[3mm] \dfrac{(\mu - \text{LSL})}{3\sigma} & \text{for larger-the-better type} \end{cases} \qquad (4.11)$$

Example 4.6

Let us now compute the C_{pk} index for the three processes represented in Figure 4.4. The lower and upper specification limits are 4 and 16, respectively; the nominal size is 10.0; and the standard deviation of all three processes, σ, is 1.5. As computed earlier, the C_p indexes for all the processes were equal to 1.33.

Process A: The process mean is $\mu_A = 10.0$; hence,

$$C_{pk} = \frac{\text{Min}[(10-4),(16-10)]}{3 \times 1.5}$$

$$= 1.333$$

When the mean is equal to the nominal size, then $C_p = C_{pk}$.

Process B: The process mean is $\mu_B = 8.5$; hence,

$$C_{pk} = \frac{\text{Min}[(8.5-4),(16-8.5)]}{3 \times 1.5}$$

$$= 1.0$$

When the mean is not equal to the nominal size, then $C_p > C_{pk}$.

Process C: The process mean is $\mu_C = 14.0$; hence,

$$C_{pk} = \frac{\text{Min}[(14-4),(16-14)]}{3 \times 1.5}$$

$$= 0.444$$

As in the case of process B, when the mean is not equal to the nominal size, $C_p > C_{pk}$.

It can be seen that the C_{pk} indexes for these processes are different, because of the differences in the means, even though the standard deviations are the same.

Now we will derive an expression for the proportion of defectives, p, in terms of the C_{pk} index. As per Eq. (4.3), for nominal-the-best type of characteristics, p is given by:

$$p = \text{Prob}[X < \text{LSL}] + \text{Prob}[X > \text{USL}]$$

$$= \text{Prob}\left[Z < \frac{\text{LSL} - \mu}{\sigma}\right] + \text{Prob}\left[Z > \frac{\text{USL} - \mu}{\sigma}\right]$$

Let us first consider the processes in which the mean is closer to the lower specification limit (LSL), as in the case of process B in Figure 4.4. Here, the C_{pk} index is

$$C_{pk} = \frac{(\mu - \text{LSL})}{3\sigma}$$

which contains information regarding the distance between the lower specification limit and the mean only which affects the proportion of the defectives to the left of the lower specification limit. It contains no information about the distance between the upper specification limit and the mean which affects the proportion of defectives to the right of the upper specification limit. It is impossible, then, to derive an expression that exactly estimates the total proportion of defectives that consists of the proportion of defectives to the left of LSL as well as the proportion of defectives to the right of USL. The best approach, then, is to derive an expression that gives an upper bound for the total proportion of defectives. As the proportion of defectives to the left of LSL is larger than the proportion of defectives to the right of USL when the mean is closer to LSL, a logical upper bound for the total proportion of defectives is two times the proportion of defectives to the left of LSL. That is,

$$p \leq 2P\left[Z < \frac{LSL - \mu}{\sigma}\right]$$

when LSL is closer to the mean, and when:

$$C_{pk} = \frac{1}{3}\frac{(\mu - LSL)}{\sigma} \tag{4.12}$$

$$p \leq 2P[Z < -3C_{pk}]$$

The processes for which the mean is closer to the upper specification limit (USL) can be handled in a similar manner. Let us consider process C in Figure 4.4. The C_{pk} index for such processes is

$$C_{pk} = \frac{(USL - \mu)}{3\sigma}$$

which can be related only to the proportion of defectives to the right of USL. As in the earlier case of B with the mean closer to LSL, the best approach here also is to derive an expression for the upper bound for the total proportion of defectives. Here, the proportion of defectives to the right of USL is larger than the proportion of defectives to the left of LSL, thus a valid upper bound for the total proportion of defectives is given by:

$$p \leq 2P\left[Z > \frac{USL - \mu}{\sigma}\right]$$

when:

$$C_{pk} = \frac{1}{3}\frac{(USL - \mu)}{\sigma} \tag{4.13}$$

$$p \leq 2P[Z > 3C_{pk}]$$

which is the same as Eq. (4.12). Therefore, the general formula for the total proportion of defectives is

$$p \leq 2P[Z > 3C_{pk}] = 2P[Z < -3C_{pk}] \tag{4.14}$$

The only assumption required in order for Eq. (4.14) to be valid is the normality of the distribution of the characteristic.

Example 4.7

Let us now estimate the proportion of defectives for processes A, B, and C in Figure 4.4 using the respective C_{pk} values and compare these estimates with the true proportion of defectives computed earlier using the means and standard deviations of these processes.

Process A: The C_{pk} index was earlier calculated in Example 4.6 as 1.333.

$$p \leq 2P[Z > 3 \times 1.33] = 2P[Z > 4.0]$$
$$= 2 \times 0.0000317 \quad \text{(from Table A.5 in Appendix)}$$
$$= 0.0000634 \ (63.4 \ \text{ppm})$$

This is the same as the estimate obtained from the C_p index and the true value obtained earlier using the mean and the standard deviation. When the mean is equal to the nominal value, the C_{pk} and C_p indexes are equal and the estimates of the proportion of defectives obtained from these indexes will be equal to the true value. (The upper bound on the total proportion of defectives computed using the C_{pk} index is the same as the exact value.)

Process B: The C_{pk} index from Example 4.6 is 1.00.

$$p \leq 2P[Z > 3 \times 1.00] = 2P[Z > 3.0]$$
$$= 2 \times 0.00135 = 0.0027 \ (2700 \ \text{ppm})$$

The exact total proportion of defectives estimated earlier using the mean and standard deviation was 0.00135 (1350 ppm). The upper bound obtained using the C_{pk} index is twice as much as the exact value.

Process C: The C_{pk} index from Example 4.6 is 0.444.

$$p \leq 2P[Z > 3 \times 0.444] = 2P[Z > 1.332]$$
$$= 2 \times 0.0918 = 0.1836 \ (183,600 \ \text{ppm})$$

The exact total proportion of defectives estimated earlier using the mean and standard deviation was 0.0918 (91,800 ppm). The upper bound obtained using the C_{pk} index is twice as much as the exact value.

It can be seen from this example that the upper bound estimated from the C_{pk} index overestimates the total proportion of defectives when the mean is not equal to the nominal size. As the C_p and C_{pk} indexes are the same for characteristics with one specification limit (smaller-the-better and larger-the-better), the proportion of defectives for these are computed in the same manner as using the C_p index (Eq. (4.9)). That is,

$$p = P[Z < -3C_{pk}] \quad \text{or} \quad P[Z > 3C_{pk}] \quad (4.15)$$

Estimates of the total percentage of defectives (and parts per million) for nominal-the-best, smaller-the-better, and larger-the-better types of characteristics from the C_p and C_{pk} indexes are given in Tables 4.5 and 4.6. For example, when the C_p index for a nominal-the-best type of characteristic is 0.90, the exact estimate of the total percentage of defectives is 0.7000 (7000 ppm) as per Table 4.5. Of course, this requires the assumption that the mean is equal to the nominal size in addition to the normality of the distribution. If the C_{pk} index is 0.90, then the upper bound of the total percentage of defectives is also 0.7000 (7000 ppm), assuming only that the distribution is normal. Consider now a characteristic with one specification limit (upper or lower) and a C_p index of 1.1. As per earlier discussion, this is also equal to the C_{pk} index. Then the estimate of the percentage of defectives is 0.04835 (483.5 ppm), as per Table 4.6.

The main limitation of the C_{pk} index is due to the normality assumption of the characteristics. Also, for the nominal-the-better type of characteristics, the C_{pk} index yields only an upper bound for the total proportion of defectives.

TABLE 4.5

Percentage Defectives for Different C_p and C_{pk} Values for Nominal-the-Best Characteristics[a]

	Upper Bound/Exact	
C_{pk}/C_p	%Defective	ppm
0.50	13.3600	133,600
0.70	3.5800	35,800
0.90	0.7000	7,000
1.00	0.2700	2,700
1.1	0.0967	967
1.2	0.03182	318.2
1.3	0.00962	96.2
1.333	0.00634	63.4
1.4	0.00267	26.7
1.5	0.00068	6.8
1.6667	0.0000574	0.574
2.0	0.0000002	0.002

[a] These results are only true if the distribution of X is normal.

TABLE 4.6

Percentage Defectives for Different Values of C_p and C_{pk} for Smaller-the-Better and Larger-the-Better Characteristics[a]

$C_p = C_{pk}$	% Defective	ppm
0.5	6.6800	66,800
0.7	1.790	17,900
0.9	0.350	3,500
1.0	0.1350	1,350
1.1	0.04835	483.5
1.2	0.01591	159.1
1.3	0.00481	48.1
1.3333	0.00317	31.7
1.4	0.001335	13.35
1.5	0.00034	3.4
1.6667	0.0000287	0.287
2.0	0.0000001	0.001

[a] These results are only true if the distribution of X is normal.

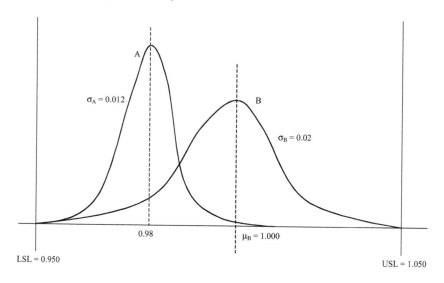

FIGURE 4.5
Processes with the same C_{pk} but different C_{pm} indexes.

As seen in Example 4.7, this could overestimate the true proportion of defectives greatly. In addition to these two obvious problems associated with the C_{pk} index, there is another flaw inherent in this index. This is best illustrated in Figure 4.5, which contains the distributions of a quality characteristic generated by two processes, A and B. The lower and upper specification limits are 0.95″ and 1.05″, respectively, thus the nominal value is 1.00″. The distributions of the characteristic generated by both the processes are normal. The distribution of process A has a mean of 0.98″ and a standard deviation of 0.012″, whereas the distribution of process B has a mean of 1.00″ (which is equal to the nominal size) and a standard deviation of 0.02". The C_p and C_{pk} indexes of these processes are as follows:

Process A:

$$C_p = \frac{(1.05 - 0.95)}{6 \times 0.012} = 1.389$$

$$C_{pk} = \frac{\text{Minimum}[(0.98 - 0.95),\ (1.05 - 0.98)]}{3 \times 0.012} = 0.833$$

Process B:

$$C_p = \frac{(1.05 - 0.95)}{6 \times 0.02} = 0.833$$

$$C_{pk} = \frac{\text{Minimum}[(1.00 - 0.95),\ (1.05 - 1.00)]}{3 \times 0.02} = 0.833$$

Process A has a smaller standard deviation compared to process B so it has a larger C_p index than process B, but this advantage of process A is compensated by a larger difference between the mean and the nominal value compared to process B, yielding the same C_{pk} index for both the processes. The C_{pk} index was introduced to take into consideration the location of the mean of the distribution. In this example, though, we see that even though the mean of process A is not equal to the target value (nominal value) and the mean of process B is equal to the target value, both have the same C_{pk} index. This is because the standard deviation of process A is less than that of process B; since the standard deviation appears in the denominator of C_{pk}, it compensates for the deviation of the mean from the target value that appears in the numerator. It is better to have an index that eliminates this problem by having both the standard deviation (or variance) and the deviation of the mean from the target value on the same side of the expression. The next index to be discussed satisfies this requirement.

4.3.4 C_{pm} Index

The C_{pm} index was developed by Chan et al. In this index for nominal-the-best characteristics, the numerator is the same as that of the C_p index, which is the range of the tolerance interval (USL − LSL). The denominator is a combined measure of the standard deviation and the deviation of the mean from the target value. The C_{pm} index is computed as:

$$C_{pm} = \frac{\text{USL} - \text{LSL}}{6\sqrt{\sigma^2 + (\mu - X_0)^2}} \tag{4.16}$$

where σ^2 is the variance of the process, μ is the mean, and X_0 is the target value. It can be seen that the expression $\sigma^2 + (\mu - X_0)^2$ in the denominator of Eq. (4.16) is the mean-squared deviation related to Taguchi's loss function discussed in Chapter 3.

Example 4.8

Compute the C_{pm} index for processes A and B in Figure 4.5.

$$LSL = 0.95"$$
$$USL = 1.05"$$
$$T = 1.00"$$
$$\mu_A = 0.98"$$
$$\sigma_A = 0.012"$$
$$\mu_B = 1.00"$$
$$\sigma_B = 0.02"$$

Process A:

$$C_{pm} = \frac{1.05 - 0.95}{6\sqrt{0.012^2 + (0.98 - 1.00)^2}}$$
$$= 0.712$$

Process B:

$$C_{pm} = \frac{1.015 - 0.95}{6\sqrt{0.020^2 + (1.00 - 1.00)^2}}$$
$$= 0.833$$

The C_{pm} index for process A is smaller than that of process B because of the larger deviation of the mean of process A from the target value. Also, the C_{pm} for process B is the same as its C_{pk} value, because its mean is equal to its target value.

The main problem with the C_{pm} index is that it cannot be used to estimate the proportion of defectives. This is because it cannot be related to the probability expression for the proportion of defectives, assuming that the characteristic is normally distributed, unlike the C_p and C_{pk} indexes. Also, it gives the same weight to the variance and the square of the deviation of the mean from the target value in the denominator. For example, the C_{pm} index for process A in Example 4.8 was less than that of process B. In some real-life applications, process A could be better than that of process B.

4.3.5 P_{pk} Index

We may recall that, in Section 4.2, wrong estimation of the mean and standard deviation was shown as one source of error in measuring the process capability.

Example 4.1 contained 30 observations collected over a period of 50 minutes. These observations were collected in six sample batches of size 5 each. The time interval between successive batches was 10 minutes. The following estimates of the process standard deviation were obtained:

1. Average value of the standard deviations of the six sample batches = 0.000738.
2. Standard deviation of the entire 30 observations taken as one sample batch = 0.001329.

It was pointed out that the estimate given by 0.000738 contains the variation within each sample batch of size 5 (short-term variability) only, whereas the estimate of 0.001329 contains the variation within the batches as well as the long-term variation in the process over a period of 50 minutes. Assuming that the process was not stopped and adjusted during the interval of 50 minutes (that is, the process control technique used to monitor the process allowed the observed deviation in the mean), the true estimate of the total variability in the characteristic is 0.001329. Usually the estimate of the variation within each batch (of size 5, in this example) is used in computing the C_p and C_{pk} indexes. As this estimate is smaller than the estimate of the total variation including the long-term variability, these indexes overestimate the process capability and hence underestimate the proportion of defectives. In order to address this problem, the P_{pk} index was introduced.

The P_{pk} index is calculated using the same formulas for computing the C_{pk} index. For nominal-the-better type of characteristics:

$$P_{pk} = \frac{\text{Minimum}[(\mu - \text{LSL}), (\text{USL} - \mu)]}{3\sigma} \tag{4.17}$$

For smaller-the-better and larger-the-better types of characteristics, the P_{pk} index is computed as:

$$P_{pk} = \frac{(\text{USL} - \mu)}{3\sigma}, \quad \text{for S-type characteristics}$$

$$= \frac{(\mu - \text{LSL})}{3\sigma}, \quad \text{for L-type characteristics} \tag{4.18}$$

In the above formulas, σ is estimated by the total long-term standard deviation. The proportion of defectives is estimated in the same manner as using the C_{pk} index. These formulas are

$$p \leq 2P[Z > 3P_{pk}] = 2P[Z < -3P_{pk}] \tag{4.19}$$

for nominal-the-best type of characteristics and

$$p = P[Z < -3P_{pk}] \quad \text{or} \quad P[Z > 3P_{pk}] \tag{4.20}$$

for smaller-the-better and larger-the-better types of characteristics. As in the case of the C_{pk} index, the distribution of the characteristic must be normal in order for Eqs. (4.19) and (4.20) to be valid.

Example 4.9

Compute the P_{pk} index for the data given in Example 4.1 and estimate the proportion of defectives using the P_{pk} index. Compare these values with the C_{pk} index and the associated estimate of the proportion of defectives. Assume that LSL = 0.995" and USL = 1.005" (see Table 4.4). Estimate 1 in Table 4.4 was obtained by averaging the individual batch estimates; the entire 30 observations considered as one batch yielded Estimate 2.

$$P_{pk} = \text{Minimum}\left\{\frac{(1.001 - 0.995), (1.005 - 1.001)}{3 \times 0.001329}\right\}$$

$$= 1.003$$
$$p \leq 2 \times P[Z > 3 \times 1.003]$$
$$= 0.0027 \ (2700 \text{ ppm})$$

The C_{pk} index for this problem will be calculated using 0.000738 as the estimate of σ in the above formula as follows:

$$C_{pk} = \text{Minimum}\left\{\frac{(1.001 - 0.995), (1.005 - 1.001)}{3 \times 0.000738}\right\}$$

$$= 1.81$$
$$p \leq 2 \times P[Z < -3 \times 1.81]$$
$$= 0.0282 \text{ ppm} \quad \text{(from Table A.5 in Appendix)}$$

It can be seen that 2700 ppm is closer to the true proportion of defectives (in ppm) rather than the 0.0282 ppm estimated from the C_{pk} index.

The limitations of the P_{pk} index are the same as those of the C_{pk} index discussed earlier. In short, these are the normality assumption required for the expressions to be valid, the upper bound on the proportion of defectives, and the masking of the deviation of the mean from the target value by the standard deviation.

4.3.6 P_p Index

The P_p index is the same as the C_p index except that it uses the estimate of the long-term standard deviation instead of the estimate of the short-term standard deviation. For a nominal-the-best type of characteristics, this is

$$P_p = \frac{(\text{USL} - \text{LSL})}{6\sigma}$$

For smaller-the-better and larger-the-better types of characteristics, the P_p index is computed as:

$$P_p = \frac{(\text{USL} - \mu)}{3\sigma}, \quad \text{for S-type characteristics}$$

$$= \frac{(\mu - \text{LSL})}{3\sigma}, \quad \text{for L-type characteristics} \tag{4.21}$$

In the above formulas, σ is estimated by the total long-term standard deviation. The proportion of defectives is estimated in the same manner as using the C_p index:

$$p = 2P[Z > 3P_p] \tag{4.22}$$

for the nominal-the-best type characteristics and

$$p = P[Z < -3P_p] \quad \text{or} \quad P[Z > 3P_p] \tag{4.23}$$

for smaller-the-better and larger-the-better types of characteristics. As in the case of the C_p index, the distribution of the characteristic must be normal and the mean must be equal to the nominal size in order for Eqs. (4.22) and (4.23) to be valid.

The limitations of the P_p index are the same as those of the C_p index discussed earlier. In short, these are the normality assumption required for the expressions to be valid and the assumption that the mean is equal to the nominal size in estimating the proportion of defectives. The P_p and P_{pk} indexes measure the performance of the process, whereas the C_p and C_{pk} indexes measure the capability of the process. For nominal-the-best type of characteristics, the C_p and P_p indexes indicate only the potential capability and performance of the process, respectively, because they do not include the process mean. The C_{pk} and P_{pk} indexes indicate the actual capability and performance of the process, respectively, because they include not only the standard deviation but also the mean.

4.4 Steps for Estimating Process Capability Indexes

This section gives the steps that must be taken to estimate the process capability of a process using one or more of the indexes discussed in the previous section. It should be noted that the major tasks in estimating the process capability are estimating properly the mean and the standard deviation of the distribution of the quality characteristic generated by the process and conducting appropriate statistical tests to ensure that the distribution of the quality characteristic is normal. The user should be aware of the sources of error and take steps to eliminate these sources as much as possible.

Step 1: Collect data. The user should collect a large amount of data randomly from the process, making certain that the process stays in "statistical control" during the entire period. Process control charts (to be discussed in Chapter 8) must be used to ensure that the process remains in control during the data-collection period. It is important to seek the customer's approval for the number of observations collected to estimate the process capability indexes.

Step 2: Analyze data. This step is necessary to check whether the assumptions required for the expressions to be valid are satisfied. This includes checking for normality and presence of outliers using statistical techniques.

Step 3: Calculate process capability indexes. The process mean, μ, and the standard deviation, σ, are estimated by the appropriate sample statistics, as discussed in Section 4.2. Customers should be informed of the sample statistics and the method used to calculate these statistics. The customer must also agree with the specification limits used in the calculation. While using the process capability indexes, it should be kept in mind that no single index value presents the complete picture and that an understanding of a process cannot be reduced to just one number. Use of the indexes must be coupled with knowledge of the technical nature of the product and processes. Also, it is important to realize that all calculated index values are estimates of the true index values because they are functions of the sample statistics, which themselves are estimates of the true values of the process mean and standard deviation. The variation in the estimates of the indexes can be reduced by taking appropriate steps while estimating the process parameters, as explained earlier in this chapter.

4.5   Estimators of Process Capability Indexes

In this section, we will study the estimation of the process capability indexes. These indexes have to be estimated because the parameters of the probability distribution of the quality characteristic required for the computation of these indexes have to be estimated. We will restrict our discussion to the nominal-the-best type of quality characteristic. The extensions to smaller-the-better and larger-the-better types of characteristics are straightforward.

4.5.1 Process Capability Ratio

$$PCR = \frac{USL - LSL}{6\sigma}$$

In this expression, only the standard deviation, σ, needs to be estimated. As we saw earlier, the unbiased estimate of the standard deviation σ is s/c_4, where s is the sample standard deviation and c_4 is a constant obtained from Table A.4 in the Appendix. Hence, the estimate of PCR is

$$\hat{PCR} = \hat{C}_p = \frac{USL - LSL}{6\frac{s}{c_4}} \tag{4.24}$$

The other estimate of σ is R/d_2, which can be used for smaller sample sizes (≤ 6). The values of d_2 can be obtained from Table A.4 in the Appendix.

If the sample size (n) is large (>30), then c_4 is close to one, thus the unbiased estimate of σ can be taken as s.

4.5.2 C_{pk} Index

$$C_{pk} = \frac{\text{Minimum}[(\mu - LSL), (USL - \mu)]}{3\sigma}$$

In this expression, both μ and σ are to be estimated. Because the unbiased estimators of these parameters are $\bar{x}$ and s/c_4, respectively, the estimate of C_{pk} is

$$\hat{C}_{pk} = \text{Min}\left[\frac{(\bar{x} - LSL)}{3\frac{s}{c_4}}, \frac{(USL - \bar{x})}{3\frac{s}{c_4}}\right] \tag{4.25}$$

In Eq. (4.25), s/c_4 can be replaced by R/d_2 for smaller sample sizes (≤ 6).

4.5.3 C_{pm} Index

For nominal-the-best type of characteristics,

$$C_{pm} = \frac{(USL - LSL)}{6\sigma'}$$

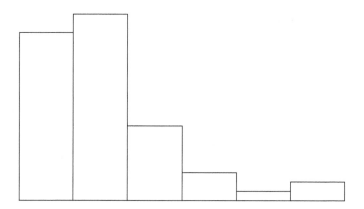

FIGURE 4.6
Histogram of C_p estimates.

where

$$\sigma' = \sqrt{\sigma^2 + (\mu - X_0)^2}$$

where X_0 is the target value for the mean. In the paper by Chan et al.,[1] the estimate of σ' is given as:

$$\hat{\sigma}' = \sqrt{\frac{\sum_{i=1}^{n}(X_i - X_0)^2}{(n-1)}} \qquad (4.26)$$

In order to illustrate the randomness of these estimators, 100 sample batches of size 10 each were simulated from a normal distribution with a known mean, $\mu = 45.0$, and standard deviation, $\sigma = 2.5$. The lower and upper specification limits (LSL and USL) were assumed to be 22.5 and 52.5, respectively, and the target value (nominal size) is $X_0 = 37.5$.

Figure 4.6 is the histogram of the 100 estimates of the C_p values obtained from 100 sample batches of size 10 each, simulated from the normal distribution with a mean equal to 45.0 and a standard deviation equal to 4.5. The estimates were computed using the following formula:

$$\hat{C}_p = \frac{(USL - LSL)}{6\left(\frac{s}{c_4}\right)}$$

The range of these estimates is (1.2264, 4.9374) and the average of these 100 values is 2.16. The true value computed using the true values of the parameters is 1.00.

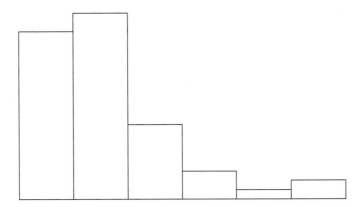

FIGURE 4.7
Histogram of C_{pk} estimates.

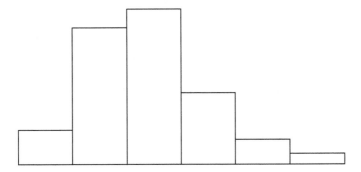

FIGURE 4.8
Histogram of C_{pm} estimates.

Figure 4.7 is the histogram of the 100 estimates of the C_{pk} index calculated using the following formula:

$$\hat{C}_{pk} = \frac{\text{Min}[(\bar{x} - \text{LSL}), (\text{USL} - \bar{x})]}{3\left(\frac{s}{c_4}\right)}$$

The range of these estimates is (0.6132, 2.4687) and the average value is 1.081. The true value is 1.00.

Figure 4.8 is the histogram of the 100 estimates of the C_{pm} index whose true value is 0.63. The range of the estimates is (0.4585, 0.8450) and the average value is 0.604. The estimates were computed using the following formula:

$$\hat{C}_{pm} = \frac{(\text{USL} - \text{LSL})}{\sqrt{\frac{\Sigma(X_i - X_0)^2}{(n-1)}}}$$

4.6 Probability Distributions of the Estimates of Process Capability Indexes

The results given in this section assume that the quality characteristic, X, is normally distributed and that the observations in the sample batch are independent. Kotz and Johnson[6] and Kotz and Lovelace[7] present excellent treatments of the process capability indexes and their estimates.

4.6.1 C_p Index

As per Kotz and Johnson,[6] the following probability statement can be made about $\hat{C}_p$ (given in Eq. (4.24)):

$$P[\hat{C}_p > c \times C_p] = P\left[\chi^2_{n-1} < \frac{(n-1)}{c^2}\right] \tag{4.27}$$

in which c is a constant, χ^2_{n-1} is the chi-square random variable with $(n-1)$ degrees of freedom (n is the sample size), and C_p is the true value of the C_p index.

The 100(1– α)% confidence interval for the true value C_p is obtained from:

$$P\left[\frac{\chi_{n-1,1-\alpha/2}}{\sqrt{n-1}}\hat{C}_p < C_p < \frac{\chi_{n-1,\alpha/2}}{\sqrt{n-1}}\hat{C}_p\right] = 1-\alpha \tag{4.28}$$

where $\chi_{n-1,\alpha}$ is the value of the square root of the chi-square random variable with $(n-1)$ degrees of freedom (n is the sample size) such that the area under the distribution to its right is "α." Chi-square values for selected values of degrees of freedom (v) and areas to the right (α) are given in Table A.6 in the Appendix. Kotz and Johnson[6] suggest the following approximations, if chi-square tables are not available:

$$\chi_{v,a} \approx \sqrt{(v-0.5)} + \frac{Z_a}{\sqrt{2}} \tag{4.29}$$

as suggested by Fisher, and

$$\chi_{v,a} \approx v^{\frac{1}{2}}\left[1 - \frac{2}{9v} + Z_a\left(\frac{2}{9v}\right)^{\frac{1}{2}}\right]^{\frac{3}{2}} \tag{4.30}$$

as suggested by Wilson and Hilferty, where v is the degrees of freedom, "a" is the area to its right, and Z_a is the value of the standard normal variable such that the area to its right is "a."

Example 4.10

A manufacturer requires that the supplier of a certain component maintain a C_p index of at least 1.50. The supplier uses a sample size of 20. What is the probability that an estimate of the C_p index ($\hat{C}_p$) obtained by the supplier will be at least 1.50?

From Eq. (4.27):

$$P[\hat{C}_p > 1.50] = P[\chi_{19}^2 < 19], \quad (c = 1 \text{ and } n = 20)$$

$$= 0.533$$

This means that the supplier could get a $\hat{C}_p$ value less than 1.50 about 46.7% $(100 - 53.3)$ of the time.

Example 4.11

The estimate of the C_p index obtained from a sample of size 15 is 1.20. Construct a 95% confidence interval for the true value of the C_p index.

In this problem, $\alpha = 0.05$ (5.0%), $(\alpha/2 = 0.025)$, $n = 15$ (degrees of freedom = 14), $Z_{0.025} = 1.96$, and $Z_{0.975} = -1.96$. From Eq. (4.29),

$$\chi_{14,0.025} \approx \sqrt{(14 - 1/2)} + \frac{1.96}{\sqrt{2}}$$

$$= 5.0602 \text{ (true value is 5.11)}$$

$$\chi_{14,0.975} \approx \sqrt{(14 - 1/2)} + -\frac{1.96}{\sqrt{2}}$$

$$= 2.2883 \text{ (true value is 2.373)}$$

From Eq. (4.28), the lower limit of the confidence interval is

$$\frac{2.2883}{\sqrt{(15 - 1)}} \times 1.20 = 0.7339$$

and the upper limit is

$$\frac{5.0602}{\sqrt{(15 - 1)}} \times 1.20 = 1.6229$$

4.6.2 C_{pk} Index

Kotz and Johnson[6] reported the following probability statement about $\hat{C}_{pk}$, the estimate of C_{pk} given in Eq. (4.25):

$$P[\hat{C}_{pk} < c] = P\left[\frac{1}{\sqrt{n}}\chi_1 + \frac{3c}{(n-1)^{1/2}}\chi_{n-1} > \frac{(USL - LSL)}{2\sigma}\right] \quad (4.31)$$

where χ_1 and χ_{n-1} are the square roots of the chi-square random variables with 1 and $(n-1)$ degrees of freedom (n is the sample size), respectively. It is very difficult to use Eq. (4.31) to make any probability statement about $\hat{C}_{pk}$, because of the presence of χ_1 and χ_{n-1} on the right-hand side. As $n \to \infty$, the distribution of $\hat{C}_{pk}$ becomes approximately equal to the distribution of $\hat{C}_p$. The 100$(1-\alpha)$% confidence interval for C_{pk} is[5,6]

$$P[\hat{C}_{pk} - E < C_{pk} < \hat{C}_{pk} + E] = (1 - \alpha) \tag{4.32}$$

where:

$$E = Z_{\alpha/2}\left\{\frac{(n-1)}{9n(n-3)} + \hat{C}_{pk}^2\frac{1}{2(n-3)}\left(1 + \frac{6}{(n-1)}\right)\right\}^{1/2} \tag{4.33}$$

where $Z_{\alpha/2}$ is the value of the standard normal random variable such that the area under its density function to its right is $\alpha/2$.

Example 4.12
The estimate of C_{pk} obtained from a sample of size 15 is 1.2. Construct a 95% confidence interval for the true C_{pk} index.

Here, $n = 15$, $\hat{C}_{pk} = 1.2$, $\alpha = 0.05$ and $Z_{\alpha/2} = Z_{0.025} = +1.96$. As per Eq. (4.33),

$$E = 1.96\left\{\frac{(15-1)}{9 \times 15 \times (15-3)} + \frac{1.2^2}{2 \times (15-3)}\left(1 + \frac{6}{(15-1)}\right)\right\}^{1/2} = 0.6021$$

The lower limit of the 95% confidence interval per Eq. (4.32) is

$$1.2 - 0.6021 = 0.5979$$

and the upper limit is

$$1.2 + 0.6021 = 1.8021$$

4.6.3 C_{pm} Index

In the paper by Chan et al.,[1] the probability density function of C_{pm} is derived. They obtained the following approximate conditional probability expression for C_{pm} using Bayes' theory. In the following expression, w is a given value:

$$P[C_{pm} > w|\hat{C}_{pm}] = P\left(Z > \left\{\frac{\left[\frac{(n-1)w}{n\hat{C}_{pm}^2}\right]^{\frac{1}{3}} - \left[1 - \frac{2}{9n}\right]}{\sqrt{\frac{2}{9n}}}\right\}\right) \tag{4.34}$$

where n is the sample size and Z is the standard normal variable.

Example 4.13

Let the LSL and USL be 0.95 and 1.05, respectively. The target value is 1.00. A sample of size 20 yielded a sample mean equal to 0.99 and a σ' equal to 0.015. Find the probability that the true value of the C_{pm} index is greater than 1.00.

$$\hat{C}_{pm} = \frac{1.05 - 0.95}{6 \times 0.015} = \frac{10}{6} = 1.11$$

$$P[C_{pm} > 1.0 \mid \hat{C}_{pm} = 1.11] = P\left(Z > \left\{ \frac{\left[\frac{(20-1) \times 1}{20 \times (1.11)^2} \right]^{\frac{1}{3}} - \left[1 - \frac{2}{9 \times 20} \right]}{\sqrt{\frac{2}{9 \times 20}}} \right\} \right)$$

$$= P[Z > -0.68] = 0.7517$$

$$P[C_{pm} < 1.0] = 1 - 0.07517 = 0.2483$$

4.7 Process Capability Indexes for Non-Normal Populations

The formulas used for computing all the process capability indexes assume that the distribution of the quality characteristic, X, is normally distributed. For example, the range 6σ or 3σ in C_p and C_{pk} assumes a certain percentage coverage of the characteristics under the normal curve (6σ range covers 99.73%). This assumption of normality is important because the calculation of the proportion of defectives (or ppm) from the capability indexes is valid only if the assumption is satisfied. English and Taylor[3] studied the robustness of the process capability indexes when the distribution of the characteristics is not normal. Somerville and Montgomery[11] give an excellent discussion of the error due to non-normality in the estimation of parts per million as the result of using equations based on the normality assumption. They recommend transformation of data (e.g., square root transformation) to convert the non-normal data to normal. It is a trial-and-error approach. After each transformation, the transformed data must be tested for normality using an appropriate statistical test. Only when the tests indicate that the transformed data are close to normal, can the formulas of the capability indexes be used. The problem with this approach is that the transformed data may not have any physical meaning and the users may not feel comfortable with the transformed data.

Many researchers have suggested using modified formulas to compute the process capability indexes after either fitting the data to the Pearson family or Johnson family of distributions, or any other appropriate distribution, and then using modified formulas.[2,4,9,10] Clements[2] suggested the following modifications for Pearson family of distributions.

Find $X_{0.00135}$ (0.135 percentile), $X_{0.50}$ (50 percentile), and $X_{0.99865}$ (99.865 percentile) and compute $\hat{C}_p$ and $\hat{C}_{pk}$ as follows for nominal-the-best type of characteristics:

$$\hat{C}_p = \frac{(USL - LSL)}{(X_{0.99865} - X_{0.00135})} \tag{4.35}$$

It can be seen that $(X_{0.99865} - X_{0.00135})$ replaces 6σ for normally distributed characteristics. Both the ranges cover the middle 99.73% of the characteristics:

$$\hat{C}_{pk} = Min\left\{ \frac{USL - X_{0.50})}{(X_{0.99865} - X_{0.50})}, \frac{(X_{0.50} - LSL)}{(X_{0.50} - X_{0.00135})} \right\} \tag{4.36}$$

In Eq. (4.36), $X_{0.50}$ replaces μ for a normal distribution. The ranges $(X_{0.99865} - X_{0.50})$ and $(X_{0.50} - X_{0.00135})$ are the intervals to the right and left of $X_{0.50}$, respectively, covering one half of 99.73% of the characteristic. Recently, McCormack et al.[8] recommended the following nonparametric indexes, based on empirical distributions.

Find $X_{0.005}$ (0.5 percentile), $X_{0.50}$ (50 percentile), and $X_{0.995}$ (99.5 percentile) and compute the following indexes:

$$\hat{C}_{np} = \frac{(USL - LSL)}{(X_{0.995} - X_{0.005})} \tag{4.37}$$

and

$$\hat{C}_{npk} = Min\left\{ \frac{(USL - X_{0.50})}{(X_{0.995} - X_{0.50})}, \frac{(X_{0.50} - LSL)}{(X_{0.50} - X_{0.005})} \right\} \tag{4.38}$$

It can be seen that the middle 99.0% coverage (instead of the conventional 99.73%) is considered in $\hat{C}_{np}$ and the left and right 49.5% coverages are considered in the $\hat{C}_{npk}$ index. These indexes are the result of the findings that in sample batches of 100 or more observations, the empirical cumulative distribution function is not very much different from the other methods, in terms of bias of extreme percentiles.

The user must be aware of the assumptions and limitations of the empirical formulas suggested by various researchers before using them.

4.8   References

1. Chan, L.K., Cheng, S.W., and Spiring, F.A., A new measure of process capability: C_{pm}, *J. Qual. Technol.*, 20, 162–175, 1988.
2. Clements, J.A., Process capability calculations for non-normal calculations, *Qual. Progr.*, 22, 95–100, 1989.

3. English, J.R. and Taylor, G.D., Process capability analysis—a robustness study, *Int. J. Prod. Res.*, 31, 1621–1635, 1993.
4. Farnum, N.R., Using Johnson curves to describe non-normal process data, *Qual. Eng.*, 9, 339–335, 1996.
5. Heavlin, W.D., *Statistical Properties of Capability Indices*, Tech. Rep. No. 320, Advanced Micro Devices, Sunnyvale, CA, 1988.
6. Kotz, S. and Johnson, N.L., *Process Capability Indices*, Chapman & Hall, London, 1993.
7. Kotz, S. and Lovelace, C.R., *Process Capability Indices in Theory and Practice*, Arnold, London, 1998.
8. McCormack, D.W., Jr., Harris, I.R., Hurwitz, A.M., and Spagon, P.D., Capability indices for non-normal data, *Qual. Eng.*, 12, 489–495, 2000.
9. Pyzdek, T., Process capability analysis using personal computers, *Qual. Eng.*, 4, 419–440, 1992.
10. Rodriguez, R.N., Recent developments in process capability analysis, *J. Qual. Technol.*, 24, 176–186, 1992.
11. Somerville, S.E. and Montgomery, D.C., Process capability indices and non-normal distributions, *Qual. Eng.*, 9, 305–316, 1996.

4.9 Problems

1. The specification limits for a nominal-the-best type of characteristic are set at 95 ± 10. The distribution of the characteristic is found to be approximately normal with a mean equal to 100.00 and a standard deviation equal to 3.00.

   a. Compute the C_p and C_{pk} indexes and the proportion of defectives using C_p and C_{pk} indexes. Compare these estimates of the proportion of defectives with the true proportion computed using a normal distribution.

   b. How much would the proportion of defectives be reduced if the mean is reduced to 95.00?

2. Consider the two processes shown below:

	Process A	Process B
Mean	100.00	105.00
Standard deviation	3.00	1.00

   Specification limits for the characteristic are set at 100 ± 3.5. Compute the C_p and C_{pk} indexes of these processes. Which process would you use? Why?

3. The tensile strength of a component has to be at least 50 ton/in.^{2}. Sample observations yield a mean of 70 tons and an estimate of standard deviation of 7.5 tons. Estimate the C_p and C_{pk} indexes and

the proportion of defectives from these values assuming a normal distribution. Compare these estimates with the estimate of the true proportion computed from a normal distribution.

4. The surface roughness of a component cannot exceed 1.01 micro-inches. The mean and standard deviation of the surface roughness are estimated to be 0.2 and 0.25, respectively. Estimate the C_p and C_{pk} indexes and the proportion of defectives using these values. Compare these estimates with the estimate of the true proportion of defectives computed from a normal distribution.

5. The following observations represent the diameters of shafts machined on a lathe:

   1.01 1.02 0.99 1.00 1.02 1.03 1.04 0.98 1.01 0.99

   The specification limits are 1 ± 0.02. Estimate the C_p and C_{pk} indexes and the proportion of defectives using the C_p and C_{pk} indexes.

6. It takes about 2 minutes for an N.C. lathe to machine the outside diameters of four pins. Sample batches, each consisting of four pins, were collected every 10 minutes. The diameters of five such sample batches are given below:

Batch #1	2.01	1.99	2.00	2.00
Batch #2	2.03	2.02	2.02	2.01
Batch #3	2.03	2.04	2.04	2.03
Batch #4	2.04	2.04	2.05	2.03
Batch #5	2.05	2.04	2.06	2.06

   Estimate the C_{pk} and P_{pk} indexes using these observations and estimate the proportion of defectives from these indexes. The specification limits are 2.02 ± 0.03.

7. The C_{pk} index of a particular batch of components is 0.9, and the C_p index is 1.2. Assume that the characteristic is nominal-the-best type and that it is normally distributed.

   a. Estimate the true proportion of defectives.

   b. If the specification are set at 6 ± 0.36, what are the possible values of the mean of the distribution?

8. The estimate of the C_p index obtained from a sample batch of size 20 is 1.3. Construct a 95% confidence interval for the true C_p index.

9. The estimate of the C_{pk} index obtained from a sample batch of size 25 is 1.5. Construct a 90% confidence interval for the true C_{pk} index.

10. The estimate of the C_{pm} index obtained from a sample batch of size 25 is 1.2. Find the probability that the true value of the C_{pm} index is greater than 1.2.

5

Measurement Error

CONTENTS

5.1 Introduction

Every measurement system has some error associated with it. The measurement error has two components—random or systematic. The random component causes a spread in the results of measurement, whereas the systematic component causes a bias in the results. The extent of bias in the measurement system is indicated by its accuracy, and the amount of variability in the measurement system is reflected in its precision. The true value of any quality characteristic is not equal to its observed value because of measurement error. There are two sources of measurement error—namely, operators and gauges. Both these sources affect both the accuracy and precision of the measurement system. Estimation of the parameters of the measurement system error is critical, because these errors affect the decisions made in process capability studies (Chapter 4), process setting (Chapter 7), process control (Chapter 8), and inspection.

5.2 Modeling the Effect of Measurement Error

Let the true value of the quality characteristic measured be X. The mean and the variance of X depend upon the process that generates X. Let these parameters be μ_x and σ_x^2, respectively. Let the observed value of the characteristic be Y, which is different from X because of the error of the measurement system.

Let the measurement error be V and its mean and variance be μ_v and σ_v^2, respectively. It is assumed that the true value X and the measurement error V are independent of each other. This assumption might be true in most real-life applications. The relationship among X, Y, and V is

$$Y = X + V \qquad (5.1)$$

Then the mean of Y is

$$\mu_y = \mu_x + \mu_v \qquad (5.2)$$

and its variance is

$$\sigma_y^2 = \sigma_x^2 + \sigma_v^2 \qquad (5.3)$$

because X and V are assumed to be independent. The variance of $Y(\sigma_y^2)$ is also called the *total variance*, and the variance of X (σ_x^2) is known as the *part-to-part variance* or the *product variance*. The mean of V, μ_v, affects the fixed component and causes a bias in the measurements. If μ_v can be estimated, then the bias can be eliminated by making suitable adjustment to the observed values such that $\mu_v = 0$. This can also be achieved by calibration of the measuring instruments, as mandated by ISO 9000 requirements. In our analysis from now on, μ_v is assumed to be 0. Then, $\mu_y = \mu_x$ as per Eq. (5.2).

The variance of V, σ_v^2, has two components—one because of operator error and the other due to gauge error. Let the variances due to operator and gauge errors be σ_o^2 and σ_g^2, respectively. Then,

$$\sigma_v^2 = \sigma_o^2 + \sigma_g^2 \qquad (5.4)$$

assuming that these error components are independent. The variance σ_o^2 is called the *variance of operator reproducibility* and the variance σ_g^2 is called the *variance of gauge repeatability*. It can be seen from Eqs. (5.3) and (5.4) that reduction of σ_o^2 and σ_g^2 will reduce σ_v^2 which in turn will bring σ_y^2 closer to σ_x^2, thereby reducing the effect of measurement error on the decisions made. The variance σ_o^2 can be reduced by training of the operators, repeated inspections, and automation, whereas the variance σ_g^2 can be reduced by proper maintenance of gauges, repeated inspections, and investment in gauges with higher precision. The estimation of σ_o^2 and σ_g^2 is done through Gauge Repeatability and Reproducibility studies,[2] described in Section 5.3.

The random variables X (true value of the quality characteristic) and V (measurement error) are assumed to be normally distributed. This assumption is true in most real-life applications. Then, the observed value is also normally distributed because of Eq. (5.1). Let the probability density functions of X, V, and Y be $f(x)$, $f(v)$, and $f(y)$, respectively. Let us also assume that X is a nominal-the-best type quality characteristic with LSL and USL as the lower and upper specification limits, respectively. This means that any component/product with a quality characteristic value within the range (LSL, USL) is accepted and any component with a characteristic value outside this range is rejected. Because of the measurement error, the decision to accept or reject

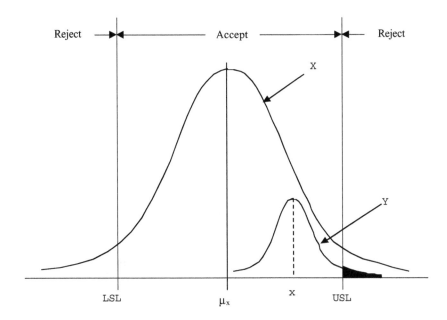

FIGURE 5.1
Rejecting a good unit.

a component/product is made based on the location of the observed value, Y, relative to the acceptance and rejection regions and is not based on the location of the true value, X. As a result, there are two types of possible wrong decisions: (1) rejecting a component/product with a true value of the characteristic falling within the acceptance region (which is a good unit), and (2) accepting a component/product with a true value of the characteristic falling outside the acceptance region (which is a defective unit). These two cases are illustrated in Figures 5.1 and 5.2.

Let us consider Figure 5.1, in which the true value of the quality characteristic, x, falls in the acceptance region. Given that the true value is x, the distribution of the observed value, Y, is normally distributed with a mean at x and a variance equal to the variance of the measurement error, V, which is σ_v^2. A part of this conditional distribution of Y given x falls in the rejection region, depending upon how close x is to the specification limits. This is represented by the shaded area in Figure 5.1, which is the conditional probability of rejecting a component, given that the true value of the characteristic is x. The probability of rejecting a good unit is the probability that Y falls in the rejection region when X is in the acceptance region. Let this probability be denoted by P_{RG}. It is derived as follows:[1]

$$P_{RG} = P[Y < LSL \text{ or } Y > USL, \quad \text{when} \quad LSL \leq X \leq USL]$$

$$= \int_{LSL}^{USL} \{P[Y < LSL] + P[Y > USL]\} f(x) \, dx \qquad (5.5)$$

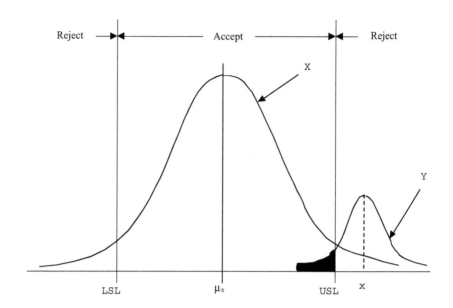

FIGURE 5.2
Accepting a defective unit.

As the conditional density function of Y is normal with a mean of x and a variance of σ_v^2, Eq. (5.5) can be written after standardization (Z is the standard normal variable) as:

$$= \int_{LSL}^{USL} \left\{ P\left[Z < \frac{LSL - x}{\sigma_v} \right] + P\left[Z > \frac{USL - x}{\sigma_v} \right] \right\} f(x)\, dx$$

$$= \int_{LSL}^{USL} \left\{ \Phi\left[\frac{LSL - x}{\sigma_v} \right] + 1 - \Phi\left[\frac{USL - x}{\sigma_v} \right] \right\} f(x)\, dx$$

$$= \int_{LSL}^{USL} f(x)\, dx - \left[\int_{LSL}^{USL} \left\{ \Phi\left[\frac{USL - x}{\sigma_v} \right] - \Phi\left[\frac{LSL - x}{\sigma_v} \right] \right\} f(x)\, dx \right] \qquad (5.6)$$

In Eq. (5.6), Φ [ ] is the cumulative probability of standard normal variable Z at the value enclosed within brackets. It can be seen that as σ_v decreases (that is, as the precision of the measurement system improves), $\{\Phi[(USL - x)/\sigma_v] - \Phi[(LSL - x)/\sigma_v]\}$ in Eq. (5.6) approaches one when x is in the interval (LSL, USL). This will cause $[\int_{LSL}^{USL} \{\Phi[(USL - x)/\sigma_v] - \Phi[(LSL - x)/\sigma_v]\} f(x)\, dx]$ to approach $\int_{LSL}^{USL} f(x)\, dx$, and the probability of rejecting a good unit, P_{RG}, will approach zero, which is the target value.

The probability of accepting a good unit, denoted by P_{AG}, is the difference between the probability that the true value, X, falls in the acceptance region and the probability of rejecting a good unit, P_{RG}, given in Eq. (5.6), and is equal to

$$P_{AG} = \int_{LSL}^{USL} f(x)\,dx$$

$$- \left(\int_{LSL}^{USL} f(x)\,dx - \left[\int_{LSL}^{USL} \left\{ \Phi\left[\frac{USL - x}{\sigma_v}\right] - \Phi\left[\frac{LSL - x}{\sigma_v}\right] \right\} f(x)\,dx \right] \right)$$

$$= \int_{LSL}^{USL} \left\{ \Phi\left[\frac{USL - x}{\sigma_v}\right] - \Phi\left[\frac{LSL - x}{\sigma_v}\right] \right\} f(x)\,dx \qquad (5.7)$$

It can be seen that as σ_v decreases (that is, as the precision of the measurement system improves), $[\int_{LSL}^{USL} \{\Phi[(USL - x)/\sigma_v] - \Phi[(LSL - x)/\sigma_v]\}f(x)dx]$ approaches $\int_{LSL}^{USL} f(x)dx$, and the probability of accepting a good unit, P_{AG}, based on the observed value, Y, will approach the probability of accepting a good unit based on the true value, X, which is the target value.

Now let us consider Figure 5.2 in which the true value of the quality characteristic, x, falls in the rejection region. Given that the true value is x, the distribution of the observed value, Y, is normally distributed with a mean at x and a variance equal to the variance of the measurement error, V, which is σ_v^2. A part of this conditional distribution of Y given x falls in the acceptance region, depending upon how close x is to the specification limits. This is represented by the shaded area in Figure 5.2, which is the conditional probability of accepting a component, given that the true value of the characteristic is x. The probability of accepting a defective unit is the probability that Y falls in the acceptance region when X is in the rejection region. Let this probability be denoted by P_{AD}. It is derived as follows:

$$P_{AD} = P[LSL \leq Y \leq USL, \text{ when } X < LSL \text{ or } X > USL]$$

$$= \int_{-\infty}^{LSL} \{P[LSL \leq Y \leq USL]\} f(x)\,dx + \int_{USL}^{\infty} \{P[LSL \leq Y \leq USL]\} f(x)\,dx$$

$$= \int_{-\infty}^{LSL} \left\{ P\left[\frac{LSL - x}{\sigma_v} \leq Z \leq \frac{USL - x}{\sigma_v}\right] \right\} f(x)\,dx$$

$$+ \int_{USL}^{\infty} \left\{ P\left[\frac{LSL - x}{\sigma_v} \leq Z \leq \frac{USL - x}{\sigma_v}\right] \right\} f(x)\,dx$$

$$= \int_{-\infty}^{LSL} \left\{ \Phi\left[\frac{USL - x}{\sigma_v}\right] - \Phi\left[\frac{LSL - x}{\sigma_v}\right] \right\} f(x)\,dx$$

$$+ \int_{USL}^{\infty} \left\{ \Phi\left[\frac{USL - x}{\sigma_v}\right] - \Phi\left[\frac{LSL - x}{\sigma_v}\right] \right\} f(x)\,dx \qquad (5.8)$$

It can be seen that as σ_v decreases (that is, as the precision of the measurement system improves), $\{\Phi[(USL - x)/\sigma_v] - \Phi[(LSL - x)/\sigma_v]\}$ in Eq. (5.8) approaches zero when x is in the interval $(-\infty, \text{LSL})$ or (USL, ∞). This will cause the probability of accepting a bad unit, P_{AD}, to approach zero, which is the target value.

The probability of rejecting a defective unit, denoted by P_{RD}, is the difference between the probability that the true value, X, falls in the rejection region and the probability of accepting a defective unit, P_{AD}, given in Eq. (5.8), and is equal to

$$P_{RD} = \int_{-\infty}^{\text{LSL}} f(x)\,dx + \int_{\text{USL}}^{\infty} f(x)\,dx - P_{AD} \qquad (5.9)$$

It can be seen that as σ_v decreases, P_{AD} approaches zero, and the probability of rejecting a defective unit, P_{RD}, based on the observed value, Y, will approach the probability of rejecting a defective unit based on the true value, X, which is the target value.

Similar probabilities of correct and wrong decisions in estimation of process capability indexes and process control can be derived. From the above analyses, it is clear that reduction of σ_v^2, the variance of measurement error, will increase the probabilities of making correct decisions. The estimation of the components of σ_v^2 (σ_o^2, variance of operator reproducibility, and σ_g^2, variance of gauge repeatability) will be described next.

5.3   Estimation of the Variance Components of Measurement Error

The estimation of σ_o^2, variance of operator reproducibility, and σ_g^2, variance of gauge repeatability, is done by conducting Gauge Repeatability and Reproducibility studies (Gauge R&R studies), described in detail in the *Measurement Systems Analysis* publication of QS-9000.[2] The recommended procedure is as follows:

1. Collect a sample batch containing n parts that represent the actual or expected range of process variation.

2. Select k operators (inspectors or appraisers), and let each operator measure the n parts in a random order and record the measurements. The gauge selected must have a discrimination of at least one tenth of the process variation. That is, if the expected process variation is 0.001", then the least count of the gauge should be at least 0.0001".

3. Repeat the cycle using the same k operators m times, changing the order of measurement in each cycle.

The method of estimation of the variance components will be illustrated in Example 5.1.

Example 5.1

Table 5.1 contains the observations recorded in a Gauge R&R study conducted with ten parts, three operators, and two cycles. Estimate the variances of operator reproducibility and gauge repeatability.

In addition to the observations, the table also contains the averages and ranges for the measurements of each part for each operator. Let the average of all the readings taken by operator i be $\overline{\overline{X}}_i$, $i = 1, 2$, and 3. In the table, $\overline{\overline{X}}_1 = 25.133$, $\overline{\overline{X}}_2 = 25.132$, and $\overline{\overline{X}}_3 = 25.129$. Also, let the average range of the readings taken by operator i be $\overline{R}_i$. In Table 5.1, $\overline{R}_1 = 0.038$, $\overline{R}_2 = 0.036$, and $\overline{R}_3 = 0.033$. Let the average of these three ranges ($\overline{R}_1, \overline{R}_2$, and $\overline{R}_3$) be $\overline{\overline{R}}$. In Example 5.1:

$$\overline{\overline{R}} = (0.038 + 0.036 + 0.033)/3 = 0.036$$

This range represents the variation due to the gauge, and the estimate of the standard deviation of the gauge repeatability is obtained by dividing $\overline{\overline{R}}$ by the appropriate d_2, tabulated in Table A.4 in the Appendix. The sample size (n) here is 2, as each range is computed from two observations. From Table A.4, d_2 for $n = 2$ is 1.128, hence the estimate of σ_g is

$$\hat{\sigma}_g = 0.036/1.128 = 0.032$$

and the estimate of the variance of gauge repeatability is

$$\hat{\sigma}_g^2 = 0.032^2 = 0.001024$$

The range of the three $\overline{\overline{X}}_i$ represents the variation due to the operators, so the estimate of the standard deviation of the operator reproducibility is obtained by dividing the range of the three $\overline{\overline{X}}_i$ by the appropriate d_2. As this range is computed from three observations ($\overline{\overline{X}}_1$, $\overline{\overline{X}}_2$, and $\overline{\overline{X}}_3$), $n = 3$, and the value of d_2 for $n = 3$ is 1.6929. The range of $\overline{\overline{X}}_i$ is

$$25.133 - 25.129 = 0.004$$

Therefore, the estimate of σ_o is

$$\hat{\sigma}_o = 0.004/1.6929 = 0.0024$$

and the estimate of the variance of operator reproducibility is

$$\hat{\sigma}_o^2 = 0.0024^2 = 0.00000576$$

Now the estimate of the total variance of the measurement system, $\hat{\sigma}_v^2$, is obtained as:

$$\hat{\sigma}_v^2 = \hat{\sigma}_g^2 + \hat{\sigma}_o^2$$
$$= 0.001024 + 0.00000576 = 0.00102976$$

TABLE 5.1

Data for Example 5.1

Part No.	Operator 1 Measurements 1	2	X̄	Range	Operator 2 Measurements 1	2	X̄	Range	Operator 3 Measurements 1	2	X̄	Range
1	25.21	25.24	25.225	0.03	25.20	25.23	25.215	0.03	25.22	25.19	25.205	0.03
2	25.33	25.31	25.32	0.02	25.31	25.28	25.295	0.03	25.29	25.32	25.305	0.03
3	24.98	25.01	24.995	0.03	25.02	25.00	25.01	0.02	24.97	24.99	24.98	0.02
4	24.99	24.98	24.985	0.01	24.98	24.99	24.985	0.01	25.00	25.01	25.005	0.01
5	25.02	25.06	25.04	0.04	25.03	25.08	25.055	0.05	25.03	25.06	25.045	0.03
6	25.40	25.38	25.39	0.02	25.39	25.36	25.375	0.03	25.41	25.39	25.4	0.02
7	24.97	25.02	24.995	0.05	24.96	25.01	24.985	0.05	24.98	25.05	25.015	0.07
8	25.11	25.18	25.145	0.07	25.14	25.19	25.165	0.05	25.12	25.16	25.14	0.04
9	25.08	25.01	25.045	0.07	25.09	25.04	25.065	0.05	25.08	25.02	25.05	0.06
10	25.17	25.21	25.19	0.04	25.15	25.19	25.17	0.04	25.15	25.13	25.14	0.02
Averages			25.133	0.038			25.132	0.036			25.1285	0.033

and the estimate of the standard deviation of the measurement system, σ_v, is

$$\hat{\sigma}_v = \sqrt{0.00102976} = 0.0321$$

This value is used in Eqs. (5.6) through (5.9) for computing the probabilities of wrong and correct decisions.

The standard deviation of the observe value, Y, which is σ_y, can be estimated using all 60 observations in Table 5.1. The standard deviation (s) of the 60 observations which is the estimate of σ_y is

$$\hat{\sigma}_y = 0.1352$$

and the estimate of the variance of Y, which is the total variance, is

$$\hat{\sigma}_y^2 = 0.1352^2 = 0.0183$$

Now using Eq. (5.3), the estimate of the variance of X which is the part-to-part variance or the product variance is obtained as follows:

$$\hat{\sigma}_x^2 = \hat{\sigma}_y^2 - \hat{\sigma}_v^2$$
$$= 0.0183 - 0.00103 = 0.01727$$

and the estimate of the standard deviation, σ_x, is

$$\hat{\sigma}_x = \sqrt{0.01727} = 0.1314$$

5.4   Minimizing the Effect of Measurement Error

The measurement system consists of the operator and the gauge. The discrimination of a measurement system is its capability to detect and indicate even small changes of the measured quality characteristic.[2] In order for the measurement system used to have adequate discrimination capabilities so that the probabilities of making the correct decisions involving process capability, process setting, process control, and inspection are maximized, the components of its variance must be very small relative to the total variance. The following percentage values capture the magnitudes of the components of the measurement system variance (MSV), relative to the total variance:

Gauge adequecy	Gauge variance (GV) $= 100 \times \dfrac{\hat{\sigma}_g}{\hat{\sigma}_y}$	(5.10)
Operator adequecy	Operator variance (OV) $= 100 \times \dfrac{\hat{\sigma}_o}{\hat{\sigma}_y}$	(5.11)
Measurement system adequecy	MSV $= 100 \times \dfrac{\hat{\sigma}_v}{\hat{\sigma}_y}$	(5.12)

In the above relations, the denominator $\hat{\sigma}_y$ can be replaced by the tolerance range, which is the difference between the upper and lower specification limits. These percentages have to be very small, if the MSV is less than or equal to 10%, then the measurement system is considered adequate. If it is greater than 10% but less than or equal to 30%, then the system may be acceptable based on the importance of the application. If the MSV is greater than 30%, then the measurement system requires improvement.[2] If the value of the MSV is large, then GV and OV can be used to identify the particular component (operator or gauge) that is the source of larger variation, and steps must be taken to reduce the associated variance component.

In Example 5.1, the percentage values are

Gauge adequacy	$GV = 100 \times 0.032/0.1352$	$= 23.67\%$
Operator adequacy	$OV = 100 \times 0.0024/0.1352$	$= 0.02\%$
Measurement system adequacy	$MSV = 100 \times 0.0321/0.1352$	$= 23.74\%$

It can be seen that as the MSV is 23% (>10%), the measurement system requires improvement. From GV and OV, it is obvious that the gauge used is not adequate. The gauge used currently must be replaced by a new gauge with better precision (smaller variance).

If the MSV is greater than 30%, then QS-9000 recommends taking multiple, statistically independent measurement readings of the quality characteristic and using the average of these readings in place of individual measurements as a temporary measure until a permanent solution is found. As the variance of the average is smaller than the variance of individual readings, this measure increases the probability of making correct decisions. The number of multiple readings for a specified MSV can be easily found as illustrated now. Let us assume that in Example 5.1, the user wants to reduce MSV from 23.74% to 15%. Then, using Eq. (5.12),

$$\hat{\sigma}_v = \frac{MSV \times \hat{\sigma}_y}{100}$$

$$= 15 \times 0.1352/100 = 0.0203$$

whereas the current $\hat{\sigma}_v = 0.0321$. Let the number of repeated readings be m. Then,

$$\frac{0.0321}{\sqrt{m}} = 0.0203,$$

from which

$$m = (0.0321/0.0203)^2 = 2.5$$

which is rounded up to 3. Therefore taking three readings of the quality characteristic and using the average of these three readings will reduce the

MSV to 15% from 23.74%. QS-9000 emphasizes that this method is only a temporary step until permanent improvements are made on the measurement system.

5.5 Other Issues Related to a Measurement System

In addition to the bias, repeatability, and reproducibility of a measurement system, QS-9000 also emphasizes its linearity and stability. Linearity measures the difference in the bias values through the expected operating range of the gauge, and stability is the total variation in the measurements obtained with a measurement system over long period of time.

5.6 References

1. Basnet, C. and Case, K.E., The effect of measurement error on accept/reject probabilities for homogeneous products, *Qual. Eng.*, 4(3), 383–397, 1992.
2. *Measurement Systems Analysis*, QS-9000, Automotive Industry Action Group, Southfield, MI, 1998.

5.7 Problems

1. The following observations were obtained in a gauge capability study. Three parts were measured by two operators. Each part was measured three times by each operator.

Part Number	Operator 1 Measurements			Operator 2 Measurements		
	1	2	3	1	2	3
1	10	11	10	11	14	12
2	9	8	11	10	11	9
3	11	11	12	12	14	12

a. Estimate the variances due to gauge and operators.
b. Estimate the MSV. If the MSV is to be 15%, what should be the number of repetitions? Comment on the answer.

2. In a study to estimate the gauge repeatability and operator reproducibility, two operators use the same gauge to measure five parts each, repeating the measurements three times for each part. The data are shown below.

Part Number	Operator 1 Measurements			Operator 2 Measurements		
	1	2	3	1	2	3
1	5	4	5	5	5	4
2	4	4	4	4	4	5
3	5	5	4	4	4	4
4	4	4	5	3	4	4
5	5	5	5	5	5	6

a. Estimate the variances due to gauge and operators.

b. Estimate the MSV. If the MSV is to be 15%, what should be the number of repetitions? Comment on the answer.

6

Optimum Process Level

CONTENTS

6.1 Introduction

As we saw in Chapter 4, the variance of the probability distribution of the quality characteristic is a function of the process selected to generate the characteristic. The optimum value of the other important parameter of the probability distribution, the mean or the process level, is determined by the user and can be achieved by process setting before start of the manufacture. The optimum value of the process mean, denoted by μ_0 and which optimizes some objective function, can be found. It should be noted that the main emphasis of this section is the modeling of objective functions under different conditions. A user may not use any of the specific models presented in this chapter, but it is the author's intention to familiarize readers with some available models and solution procedures in this area so they can build appropriate models of interest.

6.2   Optimum Process Level for Characteristics with Both Lower and Upper Limits (Nominal-the-Best Type of Characteristics)

For these types of characteristics with large variances, it is possible that the actual range of the characteristics may fall outside the tolerance region bounded by the lower and upper specification limits. Though such processes are not the best possible candidates for generating these characteristics, the process engineer may not have any other option. In such cases, the costs due to scrap and rework can be minimized by using the optimum process mean/level. If the cost to rework or scrap due to manufacturing a product with a quality characteristic above the upper specification limit is the same as the cost incurred for reworking or scrapping a product with a quality characteristic below the lower specification limit, then it is intuitive to set the optimum process mean, μ_0, equal to the mid-point of the tolerance interval. Now we will find the optimum process mean when these costs are not equal.[13]

The following assumptions are made:

1. The quality characteristic X (diameter, length, thickness, etc.) follows the probability density function $f(x)$, with an unknown mean μ and a known standard deviation σ.

2. The minimum and maximum possible values of X are X_{min} and X_{max}, respectively.

3. The lower and upper specification limits of the tolerance region are LSL and USL, respectively.

4. The cost of rework or scrap when the quality characteristic exceeds USL is C_{USL}, and the rework or scrap cost when the quality characteristic is less than LSL is C_{LSL}.

The aim here is to find the value of the process mean μ that minimizes the total expected cost per item, which is the sum of the expected cost per item when its quality characteristic X is less than LSL and the expected cost when X is greater than USL.

That is,

$$TC = E(C_{LSL}) + E(C_{USL}) \tag{6.1}$$

where TC is the total expected cost per item, $E(C_{LSL})$ denotes the expected cost of rework or scrap per item when $X < LSL$, and $E(C_{USL})$ is the expected cost of rework or scrap per item when $X > USL$:

$$E(C_{LSL}) = C_{LSL} \times \text{probability that } X \text{ is less than LSL}$$
$$= C_{LSL} \times P[X < LSL] \tag{6.2}$$

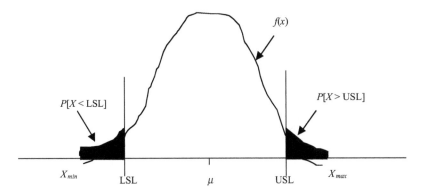

FIGURE 6.1
Acceptance and rejection probabilities.

and

$$E(C_{USL}) = C_{USL} \times \text{probability that } X \text{ is greater than USL}$$
$$= C_{USL} \times P[X > USL] \qquad (6.3)$$

The probabilities in Eqs. (6.2) and (6.3) are represented as shaded areas on the density function $f_x(x)$ in Figure 6.1.

Now, the expected costs can be written as:

$$E(C_{LSL}) = C_{LSL} \int_{X_{min}}^{LSL} f(x)\, dx \qquad (6.4)$$

and

$$E(C_{USL}) = C_{USL} \int_{USL}^{X_{max}} f(x)\, dx \qquad (6.5)$$

Hence, the total expected cost is

$$TC(\mu) = C_{LSL} \int_{X_{min}}^{LSL} f(x)\, dx + C_{USL} \int_{USL}^{X_{max}} f(x)\, dx \qquad (6.6)$$

In Eq. (6.6), the total expected cost per item is denoted by $TC(\mu)$ to signify that it is a function of the decision variable μ.

The expression in Eq. (6.6) is the objective function of our optimization problem. It consists of components $E(C_{LSL})$ and $E(C_{USL})$, out of which $E(C_{LSL})$ decreases and $E(C_{USL})$ increases as μ increases and vice versa (see Figure 6.2). This is based on the assumption that the function $TC(\mu)$ has a unique minimum value corresponding to the optimum process level μ_0. At this stage, the optimum value

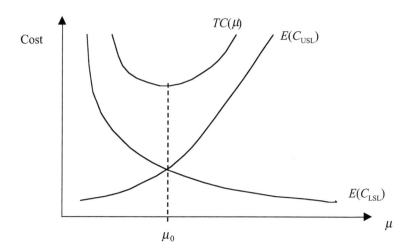

FIGURE 6.2
Relationship of costs to mean.

of μ can be obtained by increasing μ systematically from a small value and numerically evaluating $TC(\mu)$ at those values of μ until it attains its lowest value. This trial-and-error procedure will yield the global optimal solution if there is a unique optimal solution, or just one of the local optimal solutions otherwise.

It is possible to prove that this problem has a unique optimal solution and derive expressions to find that solution. Upon examining the expression for $TC(\mu)$ in Eq. (6.6), it can be seen that the decision variable μ does not appear explicitly on the right-hand side. Hence, a new variable, W, is introduced:

$$W = \frac{X - \mu}{\sigma} \tag{6.7}$$

Let the probability density function of W obtained from $f(x)$ be $g(w)$, which has a range of $(X_{min} - \mu)/\sigma, (X_{max} - \mu)/\sigma$. It can be seen that the mean and variance of W are 0 and 1, respectively. As $dx = \sigma\,dw$, Eq. (6.6) can be written as:

$$TC(\mu) = C_{LSL}\sigma\int_{(X_{min}-\mu)/\sigma}^{(LSL-\mu)/\sigma} g(w)\,dw + C_{USL}\sigma\int_{(USL-\mu)/\sigma}^{(X_{max}-\mu)/\sigma} g(w)\,dw \tag{6.8}$$

The solution procedure consists of the following steps:

1. Differentiating $TC(\mu)$ with respect to μ, setting the derivative equal to 0, and solving for μ_0
2. Proving that the optimum value of μ obtained in step (1) is the unique optimum solution that indeed minimizes $TC(\mu)$ by showing that the second derivative of $TC(\mu)$ with respect to μ is >0 at $\mu = \mu_0$.

Using the Leibniz rule, we get:

$$\frac{d(TC(\mu))}{d\mu} = \sigma C_{LSL}\left[g\left(\frac{LSL-\mu}{\sigma}\right) \times \left(-\frac{1}{\sigma}\right) - g\left(\frac{X_{min}-\mu}{\sigma}\right) \times \left(-\frac{1}{\sigma}\right)\right]$$

$$+ \sigma C_{USL}\left[g\left(\frac{X_{max}-\mu}{\sigma}\right) \times \left(-\frac{1}{\sigma}\right) - g\left(\frac{USL-\mu}{\sigma}\right) \times \left(-\frac{1}{\sigma}\right)\right] = 0$$

$$C_{LSL}\left[g\left(\frac{LSL-\mu}{\sigma}\right) - g\left(\frac{X_{min}-\mu}{\sigma}\right)\right] = C_{USL}\left[g\left(\frac{USL-\mu}{\sigma}\right) - g\left(\frac{X_{max}-\mu}{\sigma}\right)\right] \quad (6.9)$$

which yields:

$$\frac{C_{LSL}}{C_{USL}} = \frac{g\left(\frac{USL-\mu}{\sigma}\right) - g\left(\frac{X_{max}-\mu}{\sigma}\right)}{g\left(\frac{LSL-\mu}{\sigma}\right) - g\left(\frac{X_{min}-\mu}{\sigma}\right)} \quad (6.10)$$

If the ordinates at X_{max} and X_{min}—that is, $g[(X_{max}-\mu)/\sigma]$ and $g[(X_{min}-\mu)/\sigma]$—are assumed to be equal to 0, then Eq. (6.10) reduces to

$$\frac{C_{LSL}}{C_{USL}} = \frac{g\left(\frac{USL-\mu}{\sigma}\right)}{g\left(\frac{LSL-\mu}{\sigma}\right)} \quad (6.11)$$

Equation (6.11) states that the total cost of rework or scrap is minimized when the mean of the distribution of the quality characteristic is located so that the ordinates of the distribution of W at the upper and lower specification limits are proportional to the ratio of the rejection costs at the lower and upper specification limits. It can be seen from both Eqs. (6.10) and (6.11) that the optimal solution does not depend upon the individual values of C_{LSL} and C_{USL} but on the ratio of these costs. In order for the solution given in Eq. (6.10) to be optimal, the second derivative of $TC(\mu)$ with respect to μ at the optimal value has to be >0. This means that:

$$C_{LSL}\left[g'\left(\frac{LSL-\mu}{\sigma}\right) - g'\left(\frac{X_{min}-\mu}{\sigma}\right)\right] - C_{USL}\left[g'\left(\frac{USL-\mu}{\sigma}\right) - g'\left(\frac{X_{max}-\mu}{\sigma}\right)\right] > 0$$

$$(6.12)$$

From Eq.(6.12),

$$\frac{C_{LSL}}{C_{USL}} > \frac{\left[g'\left(\frac{USL-\mu}{\sigma}\right) - g'\left(\frac{X_{max}-\mu}{\sigma}\right)\right]}{\left[g'\left(\frac{LSL-\mu}{\sigma}\right) - g'\left(\frac{X_{min}-\mu}{\sigma}\right)\right]} \quad (6.13)$$

Combining this with Eq. (6.10), we get the following condition for optimality:

$$\frac{\left[g'\left(\frac{LSL-\mu}{\sigma}\right)-g'\left(\frac{X_{min}-\mu}{\sigma}\right)\right]}{\left[g\left(\frac{LSL-\mu}{\sigma}\right)-g\left(\frac{X_{min}-\mu}{\sigma}\right)\right]} > \frac{\left[g'\left(\frac{USL-\mu}{\sigma}\right)-g'\left(\frac{X_{max}-\mu}{\sigma}\right)\right]}{\left[g\left(\frac{USL-\mu}{\sigma}\right)-g\left(\frac{X_{max}-\mu}{\sigma}\right)\right]} \tag{6.14}$$

If $g[(X_{min}-\mu)/\sigma] = [g((X_{max}-\mu)/\sigma)] = 0$, then Eq. (6.14) reduces to:

$$\frac{\left[g'\left(\frac{LSL-\mu}{\sigma}\right)-g'\left(\frac{X_{min}-\mu}{\sigma}\right)\right]}{\left[g\left(\frac{LSL-\mu}{\sigma}\right)\right]} > \frac{\left[g'\left(\frac{USL-\mu}{\sigma}\right)-g'\left(\frac{X_{max}-\mu}{\sigma}\right)\right]}{\left[g\left(\frac{USL-\mu}{\sigma}\right)\right]} \tag{6.15}$$

Now we check the condition for optimality in Eq. (6.15) for normal and beta distributions and derive the optimal mean, μ_0, using Eq. (6.10).

6.2.1 Normal Density Function

In the case of the normal density function, $X_{min} = -\infty$ and $X_{max} = +\infty$; hence, $g[(X_{min}-\mu)/\sigma] = g[(X_{max}-\mu)/\sigma] = 0$. Therefore, condition (6.15) has to be satisfied for optimality:

$$g\left(\frac{X-\mu}{\sigma}\right) = \frac{1}{\sqrt{2\pi}}e^{-[(X-\mu)^2/2\sigma^2]} \quad \text{and} \quad g'\left(\frac{X-\mu}{\sigma}\right) = -\frac{1}{\sqrt{2\pi}}e^{-[(X-\mu)^2/2\sigma^2]}\left(\frac{X-\mu}{\sigma^2}\right)$$

Hence,

$$\frac{\left[g'\left(\frac{LSL-\mu}{\sigma}\right)-g'\left(\frac{X_{min}-\mu}{\sigma}\right)\right]}{\left[g\left(\frac{LSL-\mu}{\sigma}\right)\right]} = -\left(\frac{LSL-\mu}{\sigma^2}\right) \tag{6.16}$$

and

$$\frac{\left[g'\left(\frac{USL-\mu}{\sigma}\right)-g'\left(\frac{X_{max}-\mu}{\sigma}\right)\right]}{\left[g\left(\frac{USL-\mu}{\sigma}\right)\right]} = -\left(\frac{USL-\mu}{\sigma^2}\right) \tag{6.17}$$

As $LSL < USL$, $(LSL - \mu)/\sigma^2 < (USL - \mu)/\sigma^2$, hence, $-(LSL - \mu)/\sigma^2 > -(USL - \mu)/\sigma^2$ and the optimality condition in Eq. (6.15) is satisfied. Now, the optimal mean can be determined using Eq. (6.11):

$$\frac{C_{LSL}}{C_{USL}} = \frac{g\left(\frac{USL-\mu}{\sigma}\right)}{g\left(\frac{LSL-\mu}{\sigma}\right)}$$

For normal density function, this becomes:

$$\frac{C_{LSL}}{C_{USL}} = \frac{e^{-[(USL-\mu)^2/2\sigma^2]}}{e^{-[(LSL-\mu)^2/2\sigma^2]}} = e^{-[\{(USL-\mu)^2-(LSL-\mu)^2\}/2\sigma^2]} \tag{6.18}$$

Taking natural logarithms of both sides of Eq. (6.18) yields:

$$\ln\left(\frac{C_{LSL}}{C_{USL}}\right) = -\frac{(USL-\mu)^2-(LSL-\mu)^2}{2\sigma^2}$$

$$-2\sigma^2 \ln\left(\frac{C_{LSL}}{C_{USL}}\right) = (USL-\mu)^2 - (LSL-\mu)^2 \tag{6.19}$$

$$(USL + LSL - 2\mu)(USL - LSL) = -2\sigma^2 \ln\left(\frac{C_{LSL}}{C_{USL}}\right)$$

$$\mu_0 = \frac{\sigma^2}{(USL - LSL)}\ln\left(\frac{C_{LSL}}{C_{USL}}\right) + \frac{(USL + LSL)}{2} \tag{6.20}$$

When

$$C_{LSL}/C_{USL} = 1, \ln(C_{LSL}/C_{USL}) = 0 \quad \text{and} \quad \mu_0 = (USL + LSL)/2. \tag{6.21}$$

In Chapter 3 (Loss Function), we saw that the optimal value (target value) of the mean of a nominal-the-best characteristic is the nominal size that minimizes the external failure costs, captured in the expected loss. The optimal value in Eq. (6.20) considers only the expected costs of rejection, which are the internal failure costs. From Eq. (6.20), the amount by which the optimal mean, μ_0, is shifted on either side of the nominal size is

$$\frac{\sigma^2}{(USL - LSL)}\ln\left(\frac{C_{LSL}}{C_{USL}}\right)$$

which is a function of the process variance σ^2, the tolerance range (USL − LSL), and the ratio of the costs of rejection, C_{LSL} and C_{USL}. If $C_{LSL} < C_{USL}$, $\ln(C_{LSL}/C_{USL})$ is negative and the optimal mean will be shifted towards the lower specification limit. On the other hand, if $C_{LSL} > C_{USL}$, $\ln(C_{LSL}/C_{USL})$ is positive and the optimal mean will be shifted towards the upper specification limit.

Example 6.1

The specification limits of the quality characteristic of a component are 1" ± 0.04, which means the lower specification limit and upper specification limits are 0.96" and 1.04", respectively. The undersized and oversized costs are $10.00 and $8.00, respectively. The process standard deviation is 0.02. What is the optimal mean?

$$LSL = 0.96"$$
$$USL = 1.04"$$
$$\sigma^2 = 0.02^2$$
$$C_{LSL} = \$10.00$$
$$C_{USL} = \$8.00$$

$$\mu_0 = \frac{0.02^2}{(1.04 - 0.96)}\ln\left(\frac{10}{8}\right) + \frac{(1.04 + 0.96)}{2} = 0.014 + 1 = 1.014"$$

If the process variance is 0.01^2, then $\mu_0 = 0.0028 + 1 = 1.0028"$.

6.2.2 Beta Density Function

There are some disadvantages in using normal density function as the probability density function of some quality characteristics. These are mainly due to the stiffness of the normal function, as only two parameters (namely, the mean μ and the variance σ^2) are available to define the function. The beta distribution solves this problem to an extent, because it is defined by four parameters, thus yielding more flexibility. This makes the beta distribution a much better candidate for fitting real-life distributions of quality characteristics, as it covers a wide range of shapes, including symmetrical, rectangular, and skewed distributions. In addition, it has a finite range, unlike the normal distribution. This probability distribution was introduced in Chapter 1 with $a = X_{min}$ and $b = X_{max}$.

The beta probability density function is

$$f(x) = \frac{1}{(X_{max} - X_{min})B(\gamma,\eta)}\left[\frac{x - X_{min}}{X_{max} - X_{min}}\right]^{\gamma-1}\left[1 - \frac{x - X_{min}}{X_{max} - X_{min}}\right]^{\eta-1},$$
$$X_{min} \le x \le X_{max} \qquad (6.22)$$

In Eq. (6.22), γ and η are the shape parameters of the beta distribution. These, along with the lower and upper limits of the range of X, X_{min} and X_{max}, are the four parameters of the distribution. The function $B(\gamma, \eta)$ is called the *beta function* and is defined as:

$$B(\gamma,\eta) = \int_0^1 v^{\gamma-1}(1 - v)^{\eta-1}dv \qquad (6.23)$$

The mean and variance are functions of γ, η, X_{min}, and X_{max}. The mean is equal to

$$\mu = (X_{max} - X_{min})\left(\frac{\gamma}{\gamma + \eta}\right) + X_{min} \tag{6.24}$$

In the problem of determining the optimal mean, it is reasonable to assume that the shape and the range remain unchanged regardless of the value of the mean and that only the location of the distribution changes depending upon the optimal mean. This means that in Eq. (6.24), $(X_{max} - X_{min})$ and $\gamma/(\gamma + \eta)$ can be treated as constants. Now, the mean μ can be written as:

$$\mu = R \times K + X_{min}$$

where $R = (X_{max} - X_{min})$ and $K = \gamma/(\gamma + \eta)$.

The density function $g(w)$ is the standardized beta distribution given by:

$$g(w) = \frac{\sigma}{RB(\gamma, \eta)}\left[\frac{\sigma w + \mu - X_{min}}{R}\right]^{\gamma-1}\left[1 - \frac{\sigma w + \mu - X_{min}}{R}\right]^{\eta-1}$$

$$= K_1(\sigma w + RK)^{\gamma-1}\left(1 - \frac{\sigma w + RK}{R}\right)^{\eta-1} \tag{6.25}$$

where $K_1 = [\sigma/RB(\gamma, \eta)]$. When $\gamma = \eta = 1$, $g(w)$ is a uniform density function. When γ and η are not equal to 1, $g[(X_{min} - \mu)/\sigma] = g[(X_{max} - \mu)/\sigma] = g'[(X_{min} - \mu)/\sigma] = g'[(X_{max} - \mu)/\sigma]$ and the optimal condition in Eg. (6.14) to be satisfied reduces to

$$\frac{\left[g'\left(\frac{LSL - \mu}{\sigma}\right)\right]}{\left[g\left(\frac{LSL - \mu}{\sigma}\right)\right]} > \frac{\left[g'\left(\frac{USL - \mu}{\sigma}\right)\right]}{\left[g\left(\frac{USL - \mu}{\sigma}\right)\right]}$$

It can be shown that the above condition is equivalent to

$$\frac{\gamma - 1}{LSL - \mu + RK} - \frac{\eta - 1}{R - LSL + \mu - X_{min}} > \frac{\gamma - 1}{USL - \mu + RK} - \frac{\eta - 1}{R - USL + \mu - X_{min}}$$

which can be simplified to

$$\frac{\gamma - 1}{LSL - X_{min}} - \frac{\gamma - 1}{USL - X_{min}} > \frac{\eta - 1}{R - LSL + X_{min}} - \frac{\eta - 1}{R - USL + X_{min}}$$

which becomes:

$$(\gamma - 1)\left(\frac{(USL - LSL)}{(LSL - X_{min})(USL - X_{min})}\right) > (\eta - 1)\left(\frac{(LSL - USL)}{(X_{max} - LSL)(X_{max} - USL)}\right) \tag{6.26}$$

It can be seen that the left-hand side of Eq. (6.26) is >0 when γ is not equal to 1, whereas the right-hand side is <0 when η is not equal to 1, assuming that $X_{min} <$ LSL < USL $< X_{max}$. Hence, the optimality condition is satisfied when both γ and η are not equal to 1.

Now the optimal μ can be obtained using Eqs. (6.11) and (6.25):

$$\frac{C_{LSL}}{C_{USL}} = \frac{(USL - \mu + RK)^{\gamma-1}(1 - \frac{USL - \mu + RK}{R})^{\eta-1}}{(LSL - \mu + RK)^{\gamma-1}(1 - \frac{LSL - \mu + RK}{R})^{\eta-1}} \qquad (6.27)$$

As it is not possible to get a closed-form expression for the optimal mean, μ_0, the solution of Eq. (6.27) requires the use of numerical techniques.

Example 6.2

The specification for the diameter of a shaft is 3.00" ± 0.01". The cost of reworking an oversized shaft is $20.00, and the cost of scrapping an undersized shaft is $200.00. Find the optimal mean for the processes in which the diameters obey the following probability density function: beta density function with range $R = 0.036$", $\gamma = 4.0$, and $\eta = 2.0$.

SOLUTION
LSL = 2.99" and USL = 3.01", C_{LSL} = $200.00, C_{USL} = $20.00, and (C_{LSL}/C_{USL}) = 10.0. Using Eq. (6.27) gives:

$$\frac{(3.022 - \mu)^3(\mu - 2.986)}{(3.014 - \mu)^3(\mu - 2.978)} = 10.00$$

which yield three possible values for μ_0: 2.976, 3.008, and 3.016. As the mean has to be in the interval (2.99, 3.01), μ_0 is 3.008.

6.3 Optimum Process Level for Quality Characteristics with Only a Lower Limit (Larger-the-Better Type of Characteristics)

The objective of this section is to develop a methodology to determine the optimal mean of a quality characteristic of the larger-the-better type (L type). Material strength and miles per gallon are some examples of the L-type quality characteristics. These quality characteristics have a lower specification limit but no upper specification limit, and the higher the mean for the L-type quality characteristic, the better. However, the manufacturing cost increases when the mean increases, thus it is not possible to increase the value of an L-type quality characteristic without any bound. The challenge is to find an optimal mean for the L-type quality characteristic such that it will be at the highest level possible, while at the same time minimizing the manufacturing cost. The material presented in this section is based on the work by Carlsson.[5] The following assumptions are made:

1. The quality characteristic X has a lower specification level, denoted by L.
2. Products with value of the quality characteristic denoted by $X \geq L$ are accepted and sold at full price, f_a. Customers will pay an additional price of $c_a(x - L)$, where x is the value of X.
3. Products with value of $X < L$ are reprocessed and sold at a reduced price, f_r.
4. The quality characteristic X is normally distributed with a known variance σ^2 and an unknown mean μ, which is the decision variable.
5. The producer's manufacturing cost per item (UC) is separated into a fixed cost b and a variable cost c, which depends on the value of the quality characteristic X, which is x. Hence, the cost per item is

$$
\begin{aligned}
UC &= b + cx \\
&= b + cx - cL + cL \\
&= (b + cL) + c(x - L) \\
&= b_0 + c_0(x - L),
\end{aligned} \tag{6.28}
$$

where $b_0 = b + cL$ and $c_0 = c$.

The manufacturer's income per unit depends upon whether the product is accepted (that is, $X > L$) or rejected, reprocessed, or sold at a reduced price.

6.3.1 Accepted Product ($X > L$)

The manufacturer receives $\$f_a$ for an accepted item. As the customer is willing to pay an additional price of $\$c_a(x - L)$, the total income to the manufacturer for an accepted item is $f_a + c_a(x - L)$.

6.3.2 Rejected Product ($X < L$)

The manufacturer receives $\$f_r$ for a rejected item. It is reasonable to assume that $f_r < f_a$. The producer may have to compensate the customer for bad quality, not only by reducing the price from f_a to f_r, but also by a price reduction proportional to the deficit in quality, which is equal to $c_r(L - x)$. Hence, the total income to the manufacturer for a rejected item is $f_r - c_r(L - x)$, which is written as $f_r + c_r(x - L)$. It is assumed that $0 < c_a < c_r < c_0$.

Now the optimal value of the mean μ of X that maximizes the total expected net income per item will be obtained. The net income is the difference between the income per item and the manufacturing cost. The net income per item depends upon whether the quality characteristic is greater than or equal to or less than L:

1. $X < L$: The net income per item to the manufacturer is

$$
\begin{aligned}
\lambda_1(x) &= f_r + c_r(x - L) - [b_0 + c_0(x - L)] \\
&= (f_r - b_0) - (c_0 - c_r)(x - L)
\end{aligned}
\tag{6.29}
$$

2. $X > L$:

$$
\begin{aligned}
\lambda_2(x) &= f_a + c_a(x - L) - [b_0 + c_0(x - L)] \\
&= (f_a - b_0) - (c_0 - c_a)(x - L)
\end{aligned}
\tag{6.30}
$$

Let $a = f_a - b_0$, $r = f_r - b_0$ $(r < a)$, $g = c_0 - c_a$, and $(1 - p) = (c_0 - c_r)/(c_0 - c_a)$. So, $c_0 - c_r = (1 - p)(c_0 - c_a) = g(1 - p)$. Now, Eqs. (6.29) and (6.30) can be written as:

$$
\lambda_1(x) = r - g(1 - p)(x - L), \quad x < L
\tag{6.31}
$$

and

$$
\lambda_2(x) = a - g(x - L), \quad x > L
\tag{6.32}
$$

Now the expected value of the net income per item is

$$
\begin{aligned}
I(\mu) &= \int_{-\infty}^{L} \lambda_1(x) f(x)\, dx + \int_{L}^{\infty} \lambda_2(x) f(x)\, dx \\
&= \int_{-\infty}^{L} [r - g(1 - p)(x - L)] f(x)\, dx + \int_{L}^{\infty} [a - g(x - L)] f(x)\, dx
\end{aligned}
\tag{6.33}
$$

In Eq. (6.33), $f(x) = 1/\sqrt{2\pi}\sigma e^{-(x-\mu)^2/2\sigma^2}$, as per assumption 4, above. Let $z = (x - \mu)/\sigma$. Then, $x = \sigma z + \mu$ and $dx = \sigma\, dz$. Let the density function of z be $h(z)$, which is the standard normal density function. Now Eq. (6.33) can be written as:

$$
\begin{aligned}
I(\mu) &= \int_{-\infty}^{(L-\mu)/\sigma} [r - g(1 - p)(\sigma z + \mu - L)] \frac{1}{\sqrt{2\pi}} e^{-z^2/2}\, dz \\
&\quad + \int_{(L-\mu)/\sigma}^{\infty} [a - g(\sigma z + \mu - L)] \frac{1}{\sqrt{2\pi}} e^{-z^2/2}\, dz
\end{aligned}
\tag{6.34}
$$

Let $(\mu - L) = \delta$. Now the decision variable is δ instead of μ and Eq. (6.34) is

$$I(\delta) = \int_{-\infty}^{-(\delta/\sigma)} [r - g(1 - \rho)(\sigma z + \delta)] \frac{1}{\sqrt{2\pi}} e^{-z^2/2} \, dz$$

$$+ \int_{-(\delta/\sigma)}^{\infty} [a - g(\sigma z + \delta)] \frac{1}{\sqrt{2\pi}} e^{-z^2/2} \, dz$$

$$= r \int_{-\infty}^{-(\delta/\sigma)} \frac{1}{\sqrt{2\pi}} e^{-z^2/2} dz + a \int_{-(\delta/\sigma)}^{\infty} \frac{1}{\sqrt{2\pi}} e^{-z^2/2} \, dz$$

$$+ g\rho \int_{-\infty}^{-(\delta/\sigma)} (\sigma z + \delta) \frac{1}{\sqrt{2\pi}} e^{-z^2/2} \, dz$$

$$- g \int_{-\infty}^{-(\delta/\sigma)} (\sigma z + \delta) \frac{1}{\sqrt{2\pi}} e^{-z^2/2} \, dz$$

$$- g \int_{-(\delta/\sigma)}^{\infty} (\sigma z + \delta) \frac{1}{\sqrt{2\pi}} e^{-z^2/2} \, dz \tag{6.35}$$

$$= r\Phi\left(\frac{-\delta}{\sigma}\right) + a\left[1 - \Phi\left(\frac{-\delta}{\sigma}\right)\right] + g\rho \int_{-\infty}^{-(\delta/\sigma)} (\sigma z + \delta) h(z) \, dz$$

$$- g \int_{-\infty}^{\infty} (\sigma z + \delta) h(z) \, dz \tag{6.36}$$

where $\Phi(h) = \int_{-\infty}^{h} (1/\sqrt{2\pi}) e^{(-z^2/2)} dz$, which is the cumulative probability of a standard normal variable. In Eq. (6.36),

$$g \int_{-\infty}^{\infty} (\sigma z + \delta) h(z) \, dz = gE[\sigma z + \delta] = g\sigma E(z) + gE(\delta)$$

$$= g\delta, \text{ as } \quad E(z) = 0, \text{ and}$$

$$g\rho \int_{-\infty}^{-(\delta/\sigma)} (\sigma z + \delta) h(z) \, dz = g\rho\sigma \int_{-\infty}^{-(\delta/\sigma)} zh(z) \, dz + g\rho\delta \int_{-\infty}^{-(\delta/\sigma)} h(z) \, dz$$

$$= g\rho\sigma \int_{-\infty}^{-(\delta/\sigma)} zh(z) \, dz + g\rho\delta\Phi\left(-\frac{\delta}{\sigma}\right)$$

Also using the substitution $z^2/2 = v$ in $\int_{-\infty}^{-(\delta/\sigma)} zh(z) dz = \int_{-\infty}^{-(\delta/\sigma)} z \frac{1}{\sqrt{2\pi}} e^{-z^2/2} dz$ yields the following:

$$\int_{-\infty}^{-(\delta/\sigma)} zh(z) \, dz = \int_{\infty}^{(\delta^2/2\sigma^2)} \frac{1}{\sqrt{2\pi}} e^{-v} \, dv = -\int_{(\delta^2/2\sigma^2)}^{\infty} \frac{1}{\sqrt{2\pi}} e^{-v} \, dv$$

$$= -\frac{1}{\sqrt{2\pi}} \left[\frac{e^{-v}}{-1}\right]_{(\delta^2/2\sigma^2)}^{\infty} = -\frac{1}{\sqrt{2\pi}} e^{-(\delta^2/2\sigma^2)} = -h\left(\frac{\delta}{\sigma}\right)$$

Now Eq. (6.36) becomes:

$$I(\delta) = r\Phi\left(-\frac{\delta}{\sigma}\right) + a\left[1 - \Phi\left(-\frac{\delta}{\sigma}\right)\right] - g\rho\sigma h\left(\frac{\delta}{\sigma}\right) + g\rho\,\delta\Phi\left(-\frac{\delta}{\sigma}\right) - g\delta$$

$$= a - \Phi\left(-\frac{\delta}{\sigma}\right)[a - r - g\rho\delta] + g\rho\sigma h\left(\frac{\delta}{\sigma}\right) - g\delta \qquad (6.37)$$

Carlsson[5] introduces an income function which he calls "the expected relative net income level per item," given by:

$$I_r(\delta) = \frac{I(\delta) - r}{a - r} \qquad (6.38)$$

$$I_r(\delta) = \left[\frac{a - \Phi\left(-\frac{\delta}{\sigma}\right)[a - r - g\rho\delta] - g\delta - g\rho\sigma h\left(\frac{\delta}{\sigma}\right) - r}{(a - r)}\right]$$

$$= 1 - \Phi\left(-\frac{\delta}{\sigma}\right)\left[1 - \frac{\rho\delta k}{\sigma}\right] - \delta\frac{k}{\sigma} - \rho k h\left(\frac{\delta}{\sigma}\right), \qquad (6.39)$$

where $k = (g\sigma)/(a - r)$.

The problem now is to find δ that maximizes $I_r(\delta)$ in Eq. (6.39), which is done by setting the partial derivative of $I_r(\delta)$ with respect to 0 and solving for δ. Using the Leibniz rule,

$$\frac{d}{d\delta}[\Phi(-\delta/\sigma)] = \frac{d}{d\delta}\left[\int_{-\infty}^{-\delta/\sigma}\frac{e^{-z^2/2}}{\sqrt{2\pi}}dz\right] = -\frac{1}{\sigma}\frac{e^{-\delta^2/2\sigma^2}}{\sqrt{2\pi}}$$

Also,

$$\frac{d}{d\delta}[h(\delta/\sigma)] = \frac{d}{d\delta}\left(\frac{e^{-\delta^2/2\sigma^2}}{\sqrt{2\pi}}\right) = \frac{1}{\sqrt{2\pi}}e^{-\delta^2/2\sigma^2}\left(-\frac{2\delta}{2\sigma^2}\right) = -\frac{\delta}{\sqrt{2\pi}\sigma^2}e^{-\delta^2/2\sigma^2}$$

Hence,

$$I_r'(\delta) = \Phi(-\delta/\sigma)\left(\frac{\rho k}{\sigma}\right)\left(1 - \frac{\rho k\delta}{\sigma}\right)\frac{e^{-\delta^2/2\sigma^2}}{\sigma\sqrt{2\pi}} - \frac{k}{\sigma} + \frac{\rho k\delta e^{-\delta^2/2\sigma^2}}{\sqrt{2\pi}\sigma^2} = 0 \qquad (6.40)$$

The optimal value of δ can be obtained by solving Eq. (6.40) using numerical methods. Also, the proof of optimality has to be performed numerically.

As a special case when $\rho = 0$, Eq. (6.40) results in:

$$\frac{e^{-\delta^2/2\sigma^2}}{\sigma\sqrt{2\pi}} = \frac{k}{\sigma}$$

$$e^{-\delta^2/2\sigma^2} = k\sqrt{2\pi}$$

$$\frac{-\delta^2}{2\sigma^2} = \ln[k\sqrt{2\pi}]$$

$$\frac{\delta^2}{\sigma^2} = -2\ln[k\sqrt{2\pi}]$$

and the optimal value of δ is given by:

$$\frac{\delta_0}{\sigma} = \sqrt{-\ln(k^2 2\pi)} \tag{6.41}$$

It can be seen that the solution given in Eq. (6.41) is feasible only if $\ln[k^2 2\pi] < 0$.

Carlsson[5] reports that δ_0/σ is an approximately linearly decreasing function of $\ln(k)$. He also found it to be independent of ρ for values of $k < 0.1$ and suggests that Eq. (6.41) as an approximation formula to find the optimal value of δ_0 when $k < 0.1$. Table 6.1 contains comparison of exact values of $I_r(\delta)$ using numerical techniques and the approximate values of $I_r(\delta)$ obtained using Eq. (6.41), for some selected values of the cost and income parameters. It can be seen that the approximation works reasonably well.

Recently, Gungor and Chandra[10] solved the problem of finding the optimal mean of a larger-the-better type of characteristic using a goal programming approach. The goals are to maximize the process mean and to minimize the manufacturing cost, proportion of rejects, and the expected loss.

TABLE 6.1

Comparison of Exact and Approximate Values of δ_0^a

σ	f_r	c_0	c_a	c_r	a	r	g	ρ	δ_0 (exact)	k	δ_0 (approximate)
0.05	6	1	0.8	0.6	3	1	0.2	−1	0.16	0.005	0.15
0.05	6	1	0.8	0.8	3	1	0.2	0	0.16	0.005	0.15
0.05	3	1	0.8	1.2	3	−2	0.2	2	0.17	0.002	0.16
0.10	6	1	0.8	0.6	3	1	0.2	−1	0.28	0.010	0.27
0.50	6	1	0.8	0.6	3	1	0.2	−1	1.02	0.050	1.02
0.05	3	1	0.8	0.6	3	−2	0.2	−1	0.17	0.002	0.16
0.05	3	8	0.8	0.6	3	−2	7.2	−0.03	0.1	0.072	0.093
0.50	3	8	0.8	0.6	3	−2	7.2	−0.03	0.01	0.72	Not feasible

[a] $L = 10.0; b_0 = 5; f_a = 8.$

6.4   Optimum Process Level and Upper Limit of a Canning Problem

This section is based on the paper by Golhar and Pollock[7] which analyzes a canning problem where cans are filled with some ingredient. Because of the inherent variations present in the filling machine, the amount of ingredient filled in a can is a random variable, with its own probability density function, a mean, and a standard deviation. The manufacturer must make sure that the amount of ingredient in any can must be above a specified value. On the other hand, if the amount of ingredient in a can is too much, then the manufacturer incurs a loss, because the cans are sold at a fixed price, regardless of the actual amount of ingredient. (Price is usually based on the stated quantity on the label of a can.) Hence, there is an optimum setting for the mean of the distribution of ingredients in a can. Each can, after it is filled, will be weighed and the cans for which the ingredients fall within the specified lower limit and the optimum upper limit will be processed further and sold in the market for a fixed price. The cans for which the ingredients are either below the lower limit or above the upper limit will be reprocessed. The reprocessing involves reclaiming the ingredients, which results in some cost. The objective is to find jointly the optimum values of the mean and the upper limit which maximize the net profit per can.

We will assume that:

1. The weight of the ingredients in a can is a random variable, denoted by X. The probability density function of X is $f(x)$, which is assumed to be a normal distribution with a mean μ (decision variable) and a variance σ^2 (known).

2. The minimum weight of the contents in a can is L (given). The maximum weight is U, which is a decision variable.

3. The cost of contents is $\$C$ per unit weight. The cost of an empty can is negligible.

4. Underfilled cans ($X < L$) and overfilled cans ($X > U$) are emptied and the ingredients are reused, incurring a cost of $\$C_r$ (this could include the cost of production time lost).

5. A can with contents weighing between L and U is sold in the market for $\$A$.

The schematic diagram of this process is given in Figure 6.3.

Objective
The objective in this problem is to find the optimum values of μ and U (μ_0 and U_0) that maximize the expected profit per can.

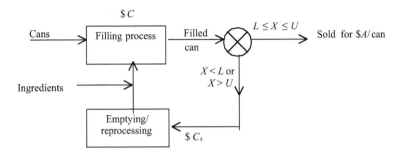

FIGURE 6.3
Schematic diagram of a canning process.

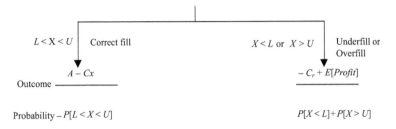

FIGURE 6.4
The outcomes of a canning problem.

Formulation of the Problem

Let the profit per can with ingredients weighing X be denoted by $P[X; \mu, U]$. We have to consider separately the cans with contents in the range (L, U) and the cans with contents less than L or greater than U. Now let us derive the expected value of the profit per can, denoted by $\bar{P}(\mu, U)$. The profit per can depends upon whether its contents are within the acceptable range or not:

1. Cans with contents in the range $L \leq X \leq U$ — As the cost/can = Cx and the income/can = A, then,

$$\text{profit/can} = A - Cx \qquad (6.42)$$

2. Cans with contents in the range $X < L$ or $X > U$ — The expected net profit per one such can equals $-C_r + E[Profit]$.

The outcomes of this problem are depicted in Figure 6.4, and now the expected profit per can be written as:

$$E[Profit\ per\ can] = E[Profit/CF]P[CF] + E[Profit/UF\ or\ OF]$$
$$\{P[UF] + P[OF]\}$$

where CF = correct fill ($L \leq X \leq U$), OF = overfill ($X > U$), and UF = underfill ($X < L$).

$E[\text{profit per can}]$

$$= \int_L^U [A - Cx] f(x) \, dx + \{-C_r + E[Profit]\}\{P[X < L] + P[X > U]\} \quad (6.43)$$

So the expected profit per can is

$$\bar{P}[\mu,U] = \int_L^U [A - Cx] f(x) \, dx + \{\bar{P}(\mu,U) - C_r\}\{P[X < L] + P[X > U]\}$$

$$= \int_L^U [A - Cx] f(x) \, dx + \{\bar{P}(\mu,U) - C_r\} p \quad (6.44)$$

where $p = 1 - P[L \leq X \leq U] = P[X < L] + P[X > U]$.
In Eq. (6.44),

$$\int_L^U [A - Cx] f(x) \, dx = A \underbrace{\int_L^U f(x) \, dx}_{1-p} - C \int_L^U x f(x) \, dx$$

$$= A(1 - p) - C \int_L^U x f(x) \, dx \quad (6.45)$$

Consider $C \int_L^U x f(x) \, dx = C \int_L^U x \dfrac{e^{-(x-\mu)^2/2\sigma^2}}{\sigma\sqrt{2\pi}} \, dx$ in Eq. (6.45). Let $Z = (X - \mu)/\sigma$.
Then,

$$C \int_L^U x \frac{e^{-(x-\mu)^2/2\sigma^2}}{\sigma\sqrt{2\pi}} \, dx = C \int_{(L-\mu)/\sigma}^{(U-\mu)/\sigma} \frac{(z+\mu) e^{-z^2/2}}{\sigma\sqrt{2\pi}} \sigma \, dz \quad (6.46)$$

Let $g(z)$ be the density function of Z which is the standard normal variable:

$$C \int_{(L-\mu)/\sigma}^{(U-\mu)/\sigma} (z+\mu) g(z) \, dz = C\sigma \int_{(L-\mu)/\sigma}^{(U-\mu)/\sigma} z g(z) \, dz + \underbrace{C\mu \int_{(L-\mu)/\sigma}^{(U-\mu)/\sigma} g(z) \, dz}_{1-p}$$

$$= C\sigma \int_{(L-\mu)/\sigma}^{(U-\mu)/\sigma} z g(z) \, dz + C\mu(1 - p) \quad (6.47)$$

Consider $\int_{(L-\mu)/\sigma}^{(U-\mu)/\sigma} z g(z)\, dz = \int_{(L-\mu)/\sigma}^{(U-\mu)/\sigma} \dfrac{z e^{-z^2/2}}{\sqrt{2\pi}}\, dz$ in Eq. (6.47), and let $z^2/2 = v$:

$$\int_{(L-\mu)/\sigma}^{(U-\mu)/\sigma} \dfrac{z e^{-z^2/2}}{\sqrt{2\pi}}\, dz = \int_{(L-\mu)/\sigma}^{0} \dfrac{z e^{-z^2/2}}{\sqrt{2\pi}}\, dz + \int_{0}^{(U-\mu)/\sigma} \dfrac{z e^{-z^2/2}}{\sqrt{2\pi}}\, dz$$

$$= \int_{(1/2)\{(L-\mu)/\sigma\}^2}^{0} \dfrac{e^{-v}\, dv}{\sqrt{2\pi}} + \int_{0}^{1/2\{(U-\mu)/\sigma\}^2} \dfrac{e^{-v}\, dv}{\sqrt{2\pi}}$$

$$= -\dfrac{1}{\sqrt{2\pi}} \int_{0}^{(1/2)\{(L-\mu)/\sigma\}^2} e^{-v}\, dv + \dfrac{1}{\sqrt{2\pi}} \int_{0}^{(1/2)\{(U-\mu)/\sigma\}^2} e^{-v}\, dv$$

$$= \dfrac{1}{\sqrt{2\pi}} [e^{-v}]_0^{(1/2)\{(L-\mu)/\sigma\}^2} - \dfrac{1}{\sqrt{2\pi}} [e^{-v}]_0^{(1/2)\{(U-\mu)/\sigma\}^2}$$

$$= \dfrac{1}{\sqrt{2\pi}} [e^{(-1/2)\{(L-\mu)/\sigma\}^2} - 1] - \dfrac{1}{\sqrt{2\pi}} [e^{(-1/2)\{(U-\mu)/\sigma\}^2} - 1]$$

$$= \dfrac{1}{\sqrt{2\pi}} [e^{(-1/2)\{(L-\mu)/\sigma\}^2} - 1 - e^{(-1/2)\{(U-\mu)/\sigma\}^2} + 1]$$

$$= -\dfrac{1}{\sqrt{2\pi}} [e^{(-1/2)\{(U-\mu)^2/\sigma^2\}} - e^{(-1/2)\{(L-\mu)^2/\sigma^2\}}]$$

$$= g\left(\dfrac{L-\mu}{\sigma}\right) - g\left(\dfrac{U-\mu}{\sigma}\right) \tag{6.48}$$

Now using Eq. (6.48) in the right-hand side of Eq. (6.47) results in:

$$C \int_{L}^{U} x \dfrac{e^{-(x-\mu)^2/2\sigma^2}}{\sigma\sqrt{2\pi}}\, dx = C\sigma\left\{ g\left(\dfrac{L-\mu}{\sigma}\right) - g\left(\dfrac{U-\mu}{\sigma}\right)\right\} + C\mu(1-p) \tag{6.49}$$

Using Eq. (6.49) in Eq. (6.46) results in:

$$\int_{L}^{U} [A - Cx] f(x)\, dx = A(1-p) - C\mu(1-p) - C\sigma\left\{ g\left(\dfrac{L-\mu}{\sigma}\right) - g\left(\dfrac{U-\mu}{\sigma}\right)\right\} \tag{6.50}$$

Using the substitution $t_1 = (U-\mu)/\sigma$ and $t_2 = (U-\mu)/\sigma$ in Eq. (6.50) gives:

$$\int_{L}^{U} [A - Cx] f(x)\, dx = A(1-p) - C\mu(1-p) - C\sigma\{ g(t_2) - g(t_1)\} \tag{6.51}$$

Now we will use Eq. (6.51) in Eq. (6.45), which then becomes:

$$\bar{P}(\mu, U) = A(1-p) - C\mu(1-p) - C\sigma\{g(t_2) - g(t_1)\}$$
$$+ \{\bar{P}(\mu, U) - C_r\}p$$

$$\bar{P}(\mu, U)[1-p] = A(1-p) - C\mu(1-p) - C\sigma\{g(t_2) - g(t_1)\} - C_r p$$

$$\bar{P}(\mu, U) = A - C\mu - \frac{C_r p}{(1-p)} - \frac{C\sigma\{g(t_2) - g(t_1)\}}{(1-p)} \tag{6.52}$$

In Eq. (6.52),

$$(1-p) = \int_{t_2}^{t_1} g(z)\,dz$$

$$= F(t_1) - F(t_2) \quad \left(F(t) = \int_{-\infty}^{t} g(z)\,dz\right)$$

and

$$-\frac{C_r p}{(1-p)} = -C_r\left[\frac{1-(1-p)}{(1-p)}\right] = \left[-\frac{C_r}{(1-p)} + C_r\right]$$

Hence, Eq. (6.52) can be written as:

$$\bar{P}(\mu, U) = A - C\mu + C_r - \frac{C_r + C\sigma\{g(t_2) - g(t_1)\}}{(1-p)} \tag{6.53}$$

Now the problem is to find the optimum values of μ and U that maximize the expected profit per can given in Eq. (6.53). It is necessary to show that $\bar{P}(\mu, U)$ is a concave function of both μ and U, which is very difficult to do analytically. It has to be done numerically. The optimal values of μ and U can be obtained by setting the partial derivatives of $\bar{P}(\mu, U)$ with respect to μ and U equal to 0 and solving the resulting equations. Golhar and Pollock[8] report the following equations to obtain the optimal values of μ and U:

$$\frac{\partial \bar{P}(\mu\mu U)}{\partial \mu} = F(t_1) - F(t_2) - f(t_2)[t_1 - t_2] = 0 \tag{6.54}$$

and

$$\frac{\partial \bar{P}(\mu\mu U)}{\partial U} = t_1\{F(t_1) - F(t_2)\} + \{f(t_1) - f(t_2)\} - \frac{C_r}{C\sigma} = 0 \tag{6.55}$$

Golhar and Pollock[7] introduce a constant $M = C_r/C\sigma$ and obtain optimum values of t_1 and t_2 using a computer program. The optimal values of μ and U are obtained from t_1 and t_2 as follows:

$$\mu_0 = L - \sigma t_{20} \tag{6.56}$$

$$U_0 = \mu_0 + \sigma t_{10} \tag{6.57}$$

where t_{10} and t_{20} are the optimal values of t_1 and t_2, respectively. The authors provide a table containing the optimal values of t_1 and t_2 for selected values of M.[7] For example, let $L = 5.0$, $C = \$0.50$, $C_r = \$0.10$, and $\sigma = 0.20$. Then $M = 0.10/(0.50 \times 0.20) = 1.0$. From the table, the optimal values of t_1 and t_2 are 1.657 and –0.750, respectively. The optimal values of μ and U are

$$\mu_0 = 5.0 - 0.10(-0.75) = 5.075$$

and

$$U_0 = 5.075 + 0.10(1.657) = 5.2407$$

The optimum mean is not located at the center of the "tolerance" interval (5.0, 5.2407). The mid-point is $(5.0 + 5.2407)/2 = 5.12035$, whereas the optimum mean is 5.075.

Let us now examine the effect of the process variability quantified in the standard deviation, σ, on the expected net profit per can. Suppose that the process does not have any variability — that is, $\sigma = 0.0$. Then, the amount of fill will be exactly equal to L and the expected net profit per can is

$$E[profit] = A - CL \tag{6.58}$$

This is the maximum profit (ideal). The expected profit $\bar{P}(\mu_0, U_0)$ for a given $\sigma > 0.0$ when $\mu = \mu_0$ and $U = U_0$ can be compared with the value given in Eq. (6.58). Let the difference between $\bar{P}(\mu_0, U_0)$ and the maximum profit given in Eq. (6.58) be dP. That is,

$$dP = A - CL - \bar{P}(\mu_0, U_0) \tag{6.59}$$

Golhar and Pollock[7] provide values of dP (denoted by $\bar{E}$ by the authors) for selected values of M, in units of $C\sigma$. For example, when $L = 5.0$, $C = \$0.50$, $C_r = \$0.10$, and $\sigma = 0.20$, $M = 1$ and $dP = 1.409$ in units of $C\sigma$, which means the reduction in the expected profit per can because of a standard deviation of 0.20 is $1.409 \times 0.50 \times 0.40 = \$0.28$. Hence, the optimal expected profit per can for the given parameter values is

$$\bar{P}(\mu_0, U_0) = (A - CL) - dP$$
$$= A - (0.5)5 - 0.28 = A - 2.78$$

Suppose that σ is halved; that is, the standard deviation is now 0.1. The new value of M is

$$M = \frac{0.1}{(0.5)(0.1)} = 2.0$$

The value of dP from the paper by Golhar and Pollock[7] for $M = 2$ is 1.663 in units of $C\sigma$. Therefore, the new expected profit per can is

$$\bar{P}(\mu_0, U_0) = A - (0.5)5 - (1.663)(0.5)(0.2) = A - 2.67$$

The net savings per can as the result of reducing the standard deviation by half is

$$\text{Savings/can} = [A - 2.67] - [A - 2.78] = 2.78 - 2.66 = \$0.11$$

This savings can be used as the basis of justifying investment in the process to improve it (reduce its variation).

Some of the researchers who have analyzed similar problems include Al-Sultan and Pulak,[1] Arcelus and Rahim,[2] Bettes,[3] Boucher and Jafari,[4] Golhar and Pollock,[7] Hunter and Kurtha,[11] and Schmidt and Pfeifer.[12]

6.5   Optimum Process Level without Considering Costs

In all the models discussed in this chapter, the objective function was either minimizing the expected cost or maximizing the expected net profit per unit. The analyses did not always yield closed-form solutions, necessitating the use of numerical methods. There are some real-life applications in which the objective is not to minimize the expected cost or to maximize the expected net profit per unit, but to satisfy one or more regulations. One such case is presented next. This problem is related to the food industry.

The manufacturers of a food product need to satisfy requirements specified by federal agencies. Two conditions to be satisfied are

1. Within a specified number of units sampled, there shall be no weight below the specified gross weight limit, which is computed as (label stated weight–specified maximum allowable variation).

2. The net average weight of a specified number of sample units must be greater than or equal to the label stated weight.

The problem is to find the optimal mean that will satisfy both these requirements. The solution procedure will be illustrated using an example. It is assumed that the weight of an individual sample, X, follows normal distribution with an unknown mean μ and a known standard deviation σ.

Example 6.3
The label stated weight (LSW) of a can is 16 ounces, and the maximum allowable variation (MAV) is 0.044 ounces. Assume that the standard deviation of the process is 0.1. Let the specified number of sample units in both requirements be ten.

REQUIREMENT 1
Assuming that the samples are independent, the number of samples out of a batch of ten with weight below (LSW – MAV) follows a binomial distribution. Let p be the probability that any one sample has a weight below (LSW – MAV). The requirement is

$$\binom{10}{0}p^0(1-p)^{10} = 1.00 \tag{6.60}$$

It can be seen that p has to be equal to 0 in order to satisfy this requirement, which means that the optimal mean of the distribution of the weight of an individual unit has to be infinity. As this is infeasible, the manufacturer has to assume a certain risk by lowering the probability on the right-hand side of Eq. (6.60). Let this be 0.90, which implies a risk of 0.10 that the requirement may not be met. Now p can be found using a modified Eq. (6.60):

$$\binom{10}{0}p^0(1-p)^{10} = 0.90$$
$$(1-p) = (0.90)^{0.10} = 0.99$$
$$p = 0.01$$

As p = probability that X is less than (LSW – MAV),

$$P[X < LSW - MAV] = 0.01$$

$$P\left[Z < \frac{(LSW - MAV) - \mu_0}{\sigma}\right] = 0.01$$

where Z is the standard normal variable. From the standard normal table,

$$P[Z < -2.33] = 0.01$$

Hence

$$\frac{(\text{LSW} - \text{MAV}) - \mu_0}{\sigma} = -2.33$$

$$\mu_0 = (\text{LSW} - \text{MAV}) + 2.33\sigma$$
$$= (16 - 0.044) + 2.33 \times 0.1 = 16.19 \text{ ounces}$$

REQUIREMENT 2

Let $\overline{X}$ be the sample mean of weights of ten sample units. Then,

$$P[\overline{X} \geq \text{LSW}] = 1.00 \qquad\qquad (6.61)$$

The sample mean $\overline{X}$ follows normal distribution with a mean μ (mean of X) and a variance of $\sigma^2/10$. It can be seen that the mean of $\overline{X}$ has to be infinity to satisfy Eq. (6.61). As in requirement 1, the manufacturer has to take a risk. Let the right-hand side of Eq. (6.61) be 0.90, resulting in a risk of 0.10 that requirement 2 may not be satisfied. Modified (6.61) is

$$P[\overline{X} \geq \text{LSW}] = 0.90$$

$$P\left[Z \geq \frac{\text{LSW} - \mu_0}{\sigma/\sqrt{10}}\right] = 0.90$$

$$p\left[Z < -\frac{\text{LSW} - \mu_0}{\sigma/\sqrt{10}}\right] = 0.10$$

From the standard normal table,

$$-\frac{\text{LSW} - \mu_0}{\sigma/\sqrt{10}} = -1.28$$

and

$$\mu_0 = \text{LSW} + 1.28\frac{\sigma}{\sqrt{10}}$$

$$= 16 + 1.28\frac{0.10}{\sqrt{10}} = 16.04$$

The optimal mean has to be 16.19 ounces in order to satisfy both requirements.

6.6 References

1. Al-Sultan, K.S. and Pulak, M.F.S., Process improvement by variance reduction for a single filling operation with rectifying inspection, *Prod. Plann. Control,* 8(5), 431–436, 1997.
2. Arcelus, F.J. and Rahim, M.A., Reducing performance variation in the canning problem, *Eur. J. Operational Res.,* 94, 477–487, 1995.
3. Bettes, D.C., Finding an optimal target in relation to a fixed lower limit and an arbitrary upper limit, *Appl. Statistics,* 11, 202–210, 1962.
4. Boucher, T.O. and Jafari, M.A., The optimum target value for single filling operations with quality sampling plan, *J. Qual. Technol.,* 23, 44–47, 1991.
5. Carlsson, O., Determining the most profitable process level for a production process under different sales conditions, *J. Qual. Technol.,* 16, 44–49, 1984.
6. Golhar, D.Y., Computation of the optimal process mean and the upper limit for a canning problem, *J. Qual. Technol.,* 20, 193–195, 1988.
7. Golhar, D.Y. and Pollock, S.M., Determination of the optimal process mean and the upper limit of the canning problem, *J. Qual. Technol.,* 20, 188–192, 1988.
8. Golhar, D.Y. and Pollock, S.M., Cost savings due to variance reduction in a canning process, *IIE Trans.,* 24, 89–92, 1992.
9. Golhar, D.Y. and Pollock, S.M., The canning problem revisited: the case of capacitated production on a fixed demand, *Eur. J. Operational Res.,* 105, 475–482, 1998.
10. Gungor, S. and Chandra, M.J., Optimal mean for a larger-the-better type characteristic using a goal programming approach, *Int. J. Ind. Eng.,* in press.
11. Hunter, W.G. and Kurtha, C.P., Determining the most profitable target value for a production process, *J. Qual. Technol.,* 9, 176–181, 1977.
12. Schmidt, R.L. and Pfeifer, P.E., An economic evaluation of improvement in process for a single-level canning problem, *J. Qual. Technol.,* 21, 16–19, 1989.
13. Springer, C.H., A method for determining the most economic position of a process mean, *Ind. Qual. Control,* 8, 36–39, 1951.

6.7 Problems

1. The specification limits of the quality characteristic of a component are 1" ± 0.005, which means the lower specification limit and upper specification limits are 0.995" and 1.005", respectively. The undersized and oversized costs are $20.00 and $60.00, respectively. The process standard deviation is 0.002. What is the optimal mean, if the quality characteristic follows a normal distribution?

2. The specification limits of the quality characteristic of a component are 20 ± 2, which means the lower specification limit and upper specification limits are 18 and 22, respectively. The undersized and oversized costs are $1.00 and $2.00, respectively. The process standard deviation is 0.50. What is the optimal mean, if the quality characteristic follows a normal distribution?

3. The specification on the diameter of a shaft are 2.00" ± 0.01". The cost of reworking an oversized shaft is $20.00, and the cost of scrapping an undersized shaft is $120.00. Find the optimal mean for the processes in which the diameters obey the following probability density function: beta density function with range $R = 0.020$", $\gamma = 3.0$, and $\eta = 1.0$.

4. The specification on the quality characteristic of a component are 3.00" ± 0.005". The cost of reworking an oversized component is $200.00, and the cost of reworking an undersized component is $10.00. Find the optimal mean for the processes in which the diameters obey the following probability density function: beta density function with range $R = 0.040$", $\gamma = 2.0$, and $\eta = 3.0$.

5. The label stated weight (LSW) of a cereal box is 8 ounces, and the maximum allowable variation (MAV) is 0.020 ounces. Also assume that the standard deviation of the process is 0.05. What should the optimal process mean be such that:

   a. Within ten units sampled, there shall be no weight below the specified gross weight limit, which is computed as the (label stated weight) – (maximum allowable variation).

   b. The net average weight of ten sample units must be greater than or equal to the label stated weight. Assume a risk of 0.05 for both requirements.

6. The cost per ounce of the contents in a soup can is $0.10, and its sales price is $2.00. The upper and lower limits are set at 16 and 15 ounces, respectively. The cost of reprocessing a can with contents outside these limits is $0.05. The process has a mean of 15.7 ounces and a standard deviation of 0.25 ounces. Assume that the distribution of contents is normal. Compute the expected profit per can earned by the manufacturer.

7. Consider the model discussed in Section 6.3 and assume the following values for the parameters:

   $L = 10.00$

   $\sigma = 0.04$

   $b_0 = \$5.00$

   $f_a = \$8.00$

   $f_r = \$6.00$

   $c_0 = 1.00$

   $c_a = \$0.80$

   $c_r = 0.60$

   Assume also that the process mean is set at 10.05 (not the optimal value) and find the expected net income per unit.

7

Process Setting

CONTENTS

7.1   Introduction

After selecting the process and improving it (σ^2 is fixed in this step at σ_0^2, which now becomes the target variance) and determining the target mean (μ_0 is fixed in this step at μ_0), the next important step is setting the process/machine so that the actual mean μ is as close as possible, if not exactly equal, to the target. Usually the operator/setter will take one or more readings on the quality characteristics and adjust the process. This is made difficult because the observations are subject to inherent variation of the process and measurement error introduced by the gauge and the operator. The discussion in this chapter is based on the work by Grubbs[2] in which he develops an optimum procedure for setting the process.

7.2   Formulation of Optimal Setting Procedure

Grubbs considered the setting process:[2]

1. The operator makes one component, measures the quality characteristic, then compares the measured value with the target value, and adjusts the process accordingly.

2. After adjustment, the operator makes another item, measures it, and adjusts the process again, if necessary.
3. These steps are repeated until no changes in setting are indicated for several consecutive items or when $\sigma/\sqrt{n}$ is suitably small, where σ is the process standard deviation and n is the number of items measured.

The following notation is used:

X = quality characteristic.

μ_0 = target value of X and of the mean of X.

σ_x^2 = variance of X.

X_i = true value of the ith item (measurement of the quality characteristic of the item made after the ith adjustment).

$\mu_i = E(X_i)$ = mean of X_i.

V = measurement error.

μ_v = mean of V which is assumed to be 0.

σ_v^2 = variance of V.

V_i = measurement error associated with ith item.

d = true deviation of μ_1 from $\mu_0 = (\mu_1 - \mu_0)$.

Y = observed value of the quality characteristic.

Y_i = observed value of the ith item = $X_i + V_i$.

k_i = factor used in ith adjustment.

The following assumptions are made by the author:

1. X and V are independent.
2. Preliminary studies have been conducted to estimate the relative measurements of the items manufactured by the process as a function of the position of the setting device (the process parameters).
3. After taking the ith measurement (Y_i), the setting is adjusted so that the mean is changed by $k_i(Y_i - \mu_0)$, where $0 < k_i < 1.0$.

Grubbs expresses the true value of the characteristic X in terms of the mean of the distribution of X and its mean μ_x as follows:

$$X = \mu + W \tag{7.1}$$

where W is a random variable with a mean = 0 and a variance = σ_x^2. Hence, the true value of the ith item is

$$X_i = \mu_i + W_i \tag{7.2}$$

The observed value of the measurement of the ith item, Y_i, is

$$Y_i = X_i + V_i = \mu_i + W_i + V_i \tag{7.3}$$

The observed value of the first item is

$$Y_1 = \mu_1 + W_1 + V_1 = \mu_0 + d + W_1 + V_1 \tag{7.4}$$

The difference between Y_1 and μ_0 is

$$Y_1 - \mu_1 = d + W_1 + V_1 \tag{7.5}$$

As an example, let us assume that the target mean μ_0 is 1" and the measurement of the first item, Y_1, is 1.2". That is,

$$Y_1 = \mu_0 + d + W_1 + V_1 = 1.2"$$

Then, the difference between Y_1 and $\mu_0 = Y_1 - \mu_0 = d + W_1 + V_1 = 0.2"$.

As per the assumption, the operator adjusts the process/machine now such that the current mean (μ_1) is reduced by $k_1(d + W_1 + V_1)$. Now the new mean is

$$
\begin{aligned}
\mu_2 &= \mu_1 - k_1(d + W_1 + V_1) \\
&= \mu_0 + d - k_1(d + W_1 + V_1) \\
&= \mu_0 + d(1 - k_1) - k_1(W_1 + V_1)
\end{aligned} \tag{7.6}
$$

Now the second observation, Y_2, is measured. This observed value is

$$
\begin{aligned}
Y_2 &= X_2 + V_2 = \mu_2 + W_2 + V_2 \\
&= \mu_0 + d(1 - k_1) - k_1(W_1 + V_1) + (W_2 + V_2)
\end{aligned} \tag{7.7}
$$

The amount of adjustment now is based on the difference between Y_2 and μ_0 which is

$$Y_2 - \mu_0 = d(1 - k_1) - k_1(W_1 + V_1) + (W_2 + V_2) \tag{7.8}$$

Now the operator adjusts the machine such that the current mean μ_2 is reduced by $k_2(Y_2 - \mu_0)$, given in Eq. (7.8). That is, the amount of adjustment is $k_2(Y_2 - \mu_0)$ and the new mean is

$$
\begin{aligned}
\mu_3 &= \mu_2 - k_2(Y_2 - \mu_0) \\
&= \mu_2 - k_2\{d(1 - k_1) - k_1(W_1 + V_1) + (W_2 + V_2)\} \\
&= \mu_0 + d(1 - k_1) - k_1(W_1 + V_1) - k_2\{d(1 - k_1) - k_1(W_1 + V_1) + (W_2 + V_2)\} \\
&= \mu_0 + d(1 - k_1)(1 - k_2) - k_1(W_1 + V_1)(1 - k_2) - k_2(W_2 + V_2)
\end{aligned} \tag{7.9}
$$

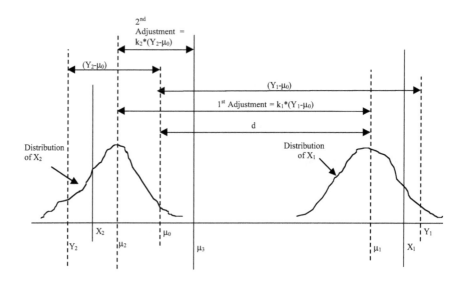

FIGURE 7.1
Process-setting procedure.

This setting procedure is illustrated in Figure 7.1.

If we proceed in a similar manner, the mean of the $(n+1)$st observation (the reading after the nth adjustment) is

$$
\begin{aligned}
\mu_{n+1} &= \mu_0 + d(1-k_1)(1-k_2)\cdots(1-k_n) \\
&\quad - k_1(1-k_2)(1-k_3)\cdots(1-k_n)(W_1+V_1) \\
&\quad - k_2(1-k_3)\cdots(1-k_n)(W_2+V_2) - k_n(W_n+V_n) \\
&= \mu_0 + d\prod_{i=1}^{n}(1-k_i) - \sum_{i=1}^{n}k_i\prod_{r=i+1}^{n}(1-k_r)(W_i+V_i) \qquad (7.10)
\end{aligned}
$$

where $\prod_{r=i+1}^{n} = 1.0$.

Now the problem is to determine the optimal values of $k_1, k_2, ..., k_n$ such that the mean after the nth adjustment is equal to μ_0 as soon as possible. A probabilistic expression of this requirement results in the following:

$$
1. \quad E(\mu_{n+1}) = \mu_0 \qquad (7.11)
$$

and

$$
2. \quad \text{Var}(\mu_{n+1}) \text{ is minimized} \qquad (7.12)
$$

Now let us examine these requirements.

CONDITION 1: $E(\mu_{n+1}) = \mu_0$
Using Eq. (7.10),

$$E\left[\mu_0 + d\prod_{i=1}^{n}(1-k_i) - \sum_{i=1}^{n}k_i \prod_{r=i+1}^{n}(1-k_r)(W_i + V_i)\right] = \mu_0 \qquad (7.13)$$

As per the assumptions made earlier, $E[W_i] = E[V_i] = 0$, for all i. Also, $E[\mu_0] = \mu_0$. Hence, Eq. (7.13) can be written as:

$$\mu_0 + d\prod_{i=1}^{n}(1-k_i) = \mu_0$$

and as d is not equal to 0,

$$\prod_{i=1}^{n}(1-k_i) = 0 \qquad (7.14)$$

CONDITION 2: $VAR[\mu_{n+1}]$ *IS MINIMIZED*

$$Var\left[\mu_0 + d\prod_{i=1}^{n}(1-k_i) - \sum_{i=1}^{n}k_i \prod_{r=i+1}^{n}(1-k_r)(W_i + V_i)\right]$$

$$= \sum_{i=1}^{n}k_i^2 \prod_{r=i+1}^{n}(1-k_r)^2[Var(W_i) + Var(V_i)]$$

$$= \sum_{i=1}^{n}k_i^2 \prod_{r=i+1}^{n}(1-k_r)^2(\sigma_x^2 + \sigma_v^2), \quad \text{as } Var(W_i) = Var(X_i) = \sigma_x^2$$

$$= (\sigma_x^2 + \sigma_v^2)\sum_{i=1}^{n}k_i^2 \prod_{r=i+1}^{n}(1-k_r)^2 \qquad (7.15)$$

Minimizing the right-hand side of Eq. (7.15) is the same as minimizing $\sum_{i=1}^{n}k_i^2\prod_{r=i+1}^{n}(1-k_r)^2$, because σ_x^2 and σ_v^2 are constants.

Now the setting problem can be stated as: Find the adjustment fractions, k_1, k_2, ..., k_n so as to minimize:

$$Z = \sum_{i=1}^{n}k_i^2 \prod_{r=i+1}^{n}(1-k_r)^2$$

subject to

$$\prod_{i=1}^{n}(1-k_i) = 0 \qquad (7.16)$$

The formulation given in Eq. (7.16) is a simple nonlinear programming problem. It can be solved using the method of LaGrange multipliers.

7.3 Solution Using LaGrange Multiplier

The unconstrained objective function of the formulation in Eq. (7.16) is

$$F(k_1, k_2, \ldots, k_n, \lambda) = \sum k_i^2 \prod_{r=i+1}^{n} (1-k_r)^2 + \lambda \left(\prod_{i=1}^{n} (1-k_i) \right) \qquad (7.17)$$

where λ is the LaGrange multiplier. Now the optimal values of $k_1, k_2, \ldots, k_n$ and λ are obtained by finding the partial derivatives $(\partial F(\cdot)/\partial k_i)$, $i = 1, 2, \ldots, n$, and $\partial F(\cdot)/\partial \lambda$ and setting them equal to 0. First let us rewrite the right-hand side of Eq. (7.17):

$$
\begin{aligned}
F(k_1, k_2, \ldots, k_n, \lambda) &= \sum_{i=1}^{n} k_i^2 \prod_{r=i+1}^{n} (1-k_r)^2 + \lambda \prod_{i=1}^{n} (1-k_i) \\
&= k_1^2 (1-k_2)^2 (1-k_3)^2 \cdots (1-k_n)^2 \\
&\quad + k_2^2 (1-k_3)^2 (1-k_4)^2 \cdots (1-k_n)^2 \\
&\quad + k_3^2 (1-k_4)^2 (1-k_5)^2 \cdots (1-k_n)^2 \\
&\quad + k_{n-1}^2 (1-k_n)^2 + k_n^2 + \lambda(1-k_1)(1-k_2) \cdots (1-k_n) \\
\frac{\partial F(\cdot)}{\partial k_1} &= 2k_1 (1-k_2)^2 \cdots (1-k_n)^2 + \lambda(-1)(1-k_2) \cdots (1-k_n) = 0
\end{aligned}
$$

from which

$$\lambda = 2k_1(1-k_2) \cdots (1-k_n) \qquad (7.18)$$

$$
\begin{aligned}
\frac{\partial F(\cdot)}{\partial k_2} &= k_1^2 2(1-k_2)(-1)(1-k_3)^2 \cdots (1-k_n)^2 + 2k_2(1-k_3)^2 \cdots (1-k_n)^2 \\
&\quad + \lambda(1-k_1)(-1)(1-k_3) \cdots (1-k_n) = 0
\end{aligned}
$$

from which

$$
\begin{aligned}
\lambda(1-k_1) + k_1[k_1 2(1-k_2) \cdots (1-k_n)] &= 2k_2(1-k_3) \cdots (1-k_n) \\
\lambda(1-k_1) + k_1(\lambda) &= 2k_2(1-k_3) \cdots (1-k_n) \\
\lambda &= 2k_2(1-k_3) \cdots (1-k_n) \qquad (7.19)
\end{aligned}
$$

By induction from Eqs. (7.18) and (7.19),

$$\lambda = 2k_i \prod_{r=i+1}^{n} (1 - k_r), \quad i = 1, 2, \ldots, n \tag{7.20}$$

Taking the partial derivative of Eq. (7.17) with respect to λ yields:

$$\frac{\partial F(\cdot)}{\partial \lambda} = \prod_{i=1}^{n} (1 - k_i) = 0 \tag{7.21}$$

In Eq. (7.21), $\prod_{i=1}^{n}(1 - k_i)$ can be shown to be equal to $1 - \Sigma_{i=1}^{n} k_i \prod_{r=i+1}^{n}(1 - k_r)$. Hence, from Eq. (7.21),

$$1 - \sum_{i=1}^{n} k_i \prod_{r=i+1}^{n} (1 - k_r) = 0$$

$$\sum_{i=1}^{n} k_i \prod_{r=i+1}^{n} (1 - k_r) = 1 \tag{7.22}$$

From Eq. (7.20), it can be seen that:

$$\sum_{i=1}^{n} k_i \prod_{r=i+1}^{n} (1 - k_r) = \sum_{i=1}^{n} \frac{\lambda}{2} = \frac{n\lambda}{2} = 1, \quad \text{and}$$

$$\lambda = \frac{2}{n} \tag{7.23}$$

Now we will find the optimum values of k_i, $i = 1, 2, 3, \ldots, n$, using Eqs. (7.20) and (7.23), starting with $i = n$:

$$2k_i \prod_{r=i+1}^{n} (1 - k_r) = 2k_n = \frac{2}{n}$$

$$k_n^* = \frac{1}{n} \tag{7.24}$$

When $i = n - 1$,

$$2k_i \prod_{r=i+1}^{n} (1 - k_r) = 2k_{n-1}(1 - k_n) = 2k_{n-1}\left(1 - \frac{1}{n}\right) = \frac{2}{n}$$

$$k_{n-1}^* = \frac{1}{n - 1} \tag{7.25}$$

By induction from Eqs. (7.24) and (7.25),

$$k_i^* = \frac{1}{i} \quad \text{for } i = 1, 2, \ldots, n \tag{7.26}$$

That is, $k_1^* = 1; k_2^* = 1/2; k_3^* = 1/3; \ldots$

It should be noted that the length of time required to converge to the target value μ_0 still depends on the variances σ_x^2 and σ_v^2. Now let us find the optimum value of the objective function in the formulation, which is the variance of μ_{n+1}, given in Eq. (7.15), using the optimal values of k_i, $i = 1, 2, \ldots, n$ given in Eq. (7.26):

$$\text{Var}(\mu_{n+1}) = (\sigma_x^2 + \sigma_v^2) \sum_{i=1}^{n} k_i^2 \prod_{r=i+1}^{n} (1-k_r)^2$$

$$\sum_{i=1}^{n} k_i^2 \prod_{r=i+1}^{n} (1-k_r)^2 = k_1^2(1-k_2)^2 \cdots (1-k_n)^2 + k_2^2(1-k_3)^2 \cdots (1-k_n)^2$$

$$+ k_3^2(1-k_4)^2 \cdots (1-k_n)^2 + \cdots + k_n^2$$

$$= 1(1-1/2)^2(1-1/3)^2 \cdots (1-1/n)^2 + (1/2)^2$$

$$\times (1-1/3)^2 \cdots (1-1/n)^2 + (1/3)^2$$

$$\times (1-1/4)^2 \cdots (1-1/n)^2 + \cdots + (1/n)^2$$

$$= 1 \times \frac{1}{2^2} \times \frac{2^2}{3^2} \times \frac{3^2}{4^2} \cdots \frac{(n-2)^2(n-1)^2}{(n-1)^2 \, n^2} + \frac{1}{2^2} \times \frac{2^2}{3^2}$$

$$\times \frac{3^2}{4^2} \cdots \frac{(n-2)^2(n-1)^2}{(n-1)^2 \, n^2} + \cdots + \frac{1}{n^2}$$

$$= n\left(\frac{1}{n^2}\right) = \frac{1}{n}$$

Hence, the optimal variance of μ_{n+1} is

$$\text{Var}(\mu_{n+1}) = \frac{(\sigma_x^2 + \sigma_v^2)}{n} \tag{7.27}$$

7.4 Multiple Work Pieces after Each Adjustment

Suppose that m work pieces are produced each time instead of just one piece and that the mean of these m measurements is used to determine the amount of setting. This means that after the $(i-1)$th adjustment, the observations are $Y_{i1}, Y_{i2}, \ldots,$ and Y_{im}. This method is compared with the setting procedure based

TABLE 7.1

Multiple Measurements After Each Adjustment

One Reading (Y_i)	m Readings $(Y_{i1}, Y_{i2}, ..., Y_{im})$: $\bar{Y}_i = \sum\limits_{j=1}^{m} \dfrac{Y_{ij}}{m}$
$Y_1 = \mu_0 + d + W_1 + V_1$	$Y_{ij} = (\mu_0 + d) + W_{1j} + V_{1j};$ $\bar{Y}_1 = (\mu_0 + d)\sum\limits_{j=1}^{m}(W_{1j} + V_{1j})$
Amount of first adjustment $\quad = k_1(Y_1 - \mu_0)$ $\quad = k_1(d + W_1 + V_1)$	$k_1(\bar{Y}_1 - \mu_0) = k_1\left[d + \dfrac{1}{m}\sum\limits_{j=1}^{m}(W_{1j} + V_{1j})\right]$
$\mu_2 = \mu_0 + d(1 - k_1) - k_1(W_1 + V_1)$	$\mu_2 = \mu_0 + d(1 - k_1) - k_1\sum\limits_{j=1}^{m}\dfrac{W_{1j} + V_{1j}}{m}$
$\mu_{n+1} = \mu_0 + \prod\limits_{i=1}^{n}d(1 - k_i)$ $\qquad - \sum\limits_{i=1}^{n}k_i\prod\limits_{r=i+1}^{n}(1 - k_r)(W_i + V_i)$	$\mu_{n+1} = \mu_0 + \prod\limits_{i=1}^{k}d(1 - k_i)$ $\qquad - \sum\limits_{i=1}^{n}k_i\prod\limits_{r=(i+1)}^{n}(1 - k_r)\left[\sum\limits_{j=1}^{m}\dfrac{(W_{ij} + V_{ij})}{m}\right]$
$E[\mu_{n+1}] = \mu_0 + d\prod\limits_{i=1}^{n}(1 - k_i)$	$E(\mu_{n+1}) = \mu_0 + \prod\limits_{i=1}^{k}d(1 - k_i)$
$\mathrm{Var}(\mu_{n+1}) = \sum\limits_{i=1}^{n}k_i^2\prod\limits_{r=i+1}^{n}(1 - k_r)(\sigma_x^2 + \sigma_v^2)$	$\mathrm{Var}(\mu_{n+1}) = \sum\limits_{i=1}^{n}k_i^2\prod\limits_{r=i+1}^{n}(1 - k_r)^2\left[\dfrac{\sigma_x^2 + \sigma_v^2}{m}\right]$
$k_i^* = \dfrac{1}{i}, i = 1, 2 \ldots, n$	$k_i^* = \dfrac{1}{i}, i = 1, 2, \ldots, n$

on one observation after each adjustment, assumed until now, in Table 7.1. Only the key steps are given. It can be seen that the optimal values of the adjustment fractions are the same as before.

7.5 Recent Developments

Trietsch[3] states that a Bayesian formulation could yield a complete solution to the adjustment problem solved by Grubbs. Recently del Castillo and Pan[1] have proposed a process-setting procedure using a Bayesian approach based on a Kalman filter estimate of d, which is the initial difference between μ_1 and the target value μ_0.

7.6 References

1. del Castillo, E. and Pan, R., A Unifying View of Some Process Adjustment Methods, working paper, Department of Industrial and Manufacturing Engineering, The Pennsylvania State University, University Park, 2000.
2. Grubbs, F.E., An optimum procedure for setting machines or adjusting processes, *Ind. Qual. Control*, July, 1954, reprinted in *J. Qual. Technol.*, 15, 186–189, 1983.
3. Trietsch, D., The harmonic rule for process setup adjustment with quadratic loss, *J. Qual. Technol.*, 30, 75–84, 1998.

7.7 Problems

1. Consider the paper by Grubbs on process setting.[2] Make the following assumptions which are different from the assumptions made by Grubbs. Assume that after each setting, the operator collects one piece from the process/machine and takes m measurements of the characteristics of the same piece. The amount of setting is based on the difference between the mean of these m readings and the target value. Derive expressions for the expected value and variance of μ_{n+1}. Obtain the optimal setting policy (that is, find the optimal values of k_i), without going through the steps of optimization. What is the variance of μ_{n+1} if the optimal setting policy is used?

2. The operator takes only one measurement as assumed by Grubbs.[2] But, the mean of the measurement error V is μ_v, which is not equal to 0 and is known. Derive expressions for the expected value and the variance of μ_{n+1}. Describe how you would modify the setting procedure developed by Grubbs to compensate for the fact that the mean of V is not 0 (that is, the instrument has a known bias).

3. Consider the paper on process adjustments by Grubbs.[2] The operator takes a sample batch of m individual pieces before each adjustment (m observations) and uses the sample mean of these m observations to make the adjustment. Make all assumptions made by Grubbs which means that the optimal adjustment policy as obtained by Grubbs can be used. In addition, make the following assumptions:

   The sampling cost consists of a fixed cost of C_f per batch and a variable cost of C_v per observation.

   The operator collects only n sample batches (that is, makes only n adjustments) before starting the regular production run. The penalty cost for starting the regular production run after n adjustments is C_a^* (variance of μ_{n+1}).

Assume that C_f, C_v, and C_a are given.

$$\text{Var}(\mu_{n+1}) = \frac{(\sigma_x^2 + \sigma_v^2)}{mn}$$

where σ_x^2 and σ_v^2 are the variances of the process and the measurement error, respectively.

a. Derive an expression for the total cost before starting regular production (after n adjustments), consisting of the total sampling cost and the penalty cost (given above), assuming that only n adjustments are made.

b. Consider m as a continuous decision variable and derive an expression for the optimal value of m that minimizes the total cost derived in a, above.

8

Process Control

CONTENTS

8.1   Introduction

Before starting manufacture of an entire lot or batch of components (or products), two conditions must be satisfied to reduce the proportion of defectives
and external failure costs:

1. The variance (σ^2) of the distribution of the characteristic should be
    minimum, which is achieved by selecting a suitable process and by
    eliminating sources of variation (using design of experiments, etc.).
2. The mean (μ) of the distribution of the characteristic should be as
    close as possible, if not equal, to the target value, which is achieved
    by process setting, including selection of the levels of process param-
    eters (using design of experiments, etc.).

Let the values of the standard deviation and the mean achieved before
starting regular manufacture be σ_0 and μ_0, respectively. These are called
"in-control" values. It is important that the mean and standard deviation of
the distribution of the quality characteristic remain equal to μ_0 and σ_0, respectively, throughout the entire duration of manufacture.

To do so requires that the process be monitored continuously during manufacture (to ensure that $\mu = \mu_0$ and $\sigma = \sigma_0$). Continuous monitoring may not
be feasible in all cases. In such situations, monitoring of the process at regular
intervals must be done. This implies that sample observations must be collected from the process at regular intervals and inferences made concerning
μ and σ. Statistically speaking, this means testing of hypotheses. There are
two sets of hypotheses that need to be tested:

1. H_0: $\mu = \mu_0$.
    H_1: $\mu \neq \mu_0$.
2. H_0: $\sigma = \sigma_0$.
    H_1: $\sigma \neq \sigma_0$.

If both null hypotheses are true ($\mu = \mu_0$ and $\sigma = \sigma_0$), then the process is said to be "in control." If either one or both are not true (false) ($\mu \neq \mu_0$, or $\sigma \neq \sigma_0$ or $\mu \neq \mu_0$ and $\sigma \neq \sigma_0$), then the process is said to be "out-of-control."

8.2 Preliminaries

8.2.1 Steps in Hypothesis Testing

1. Set up hypotheses (we'll consider only μ now):

$$H_0: \mu = \mu_0; \quad H_1: \mu \neq \mu_0$$

2. Select an unbiased statistic (with the minimum variance, if possible); ($\bar{X}$ for μ).

3. Specify probability of Type I error, which is

$$P \text{ [Rejecting } H_0, \text{ when it is true]} = \alpha$$

4. Using α and the distribution of $\bar{X}$, set up the acceptance regions for $\bar{X}$:

Accept H_0, if $\mu_0 - Z_{\alpha/2} \times \sigma_0/\sqrt{n} < \bar{X} < \mu_0 + Z_{\alpha/2} \times \sigma_0/\sqrt{n}$.

Reject H_0, if $\bar{X} < \mu_0 - Z_{\alpha/2} \times \sigma_0/\sqrt{n}$ or $\bar{X} > \mu_0 + Z_{\alpha/2} \times \sigma_0/\sqrt{n}$.

In process control, when H_0 is accepted, the process is assumed to be "in-control" and is allowed to run. If H_0 is rejected, the process is assumed to be "out-of-control" and stopped. Then the process must be examined to identify the "assignable causes" or "special causes" that might have caused the "out-of-control" state of the process. Once the assignable causes are located, they must be removed and then only the process is allowed to run. Since this procedure has to be repeated every h time units ($h = 5, 10, 20,$ or 30 minutes), it is convenient to represent this procedure by a chart. This is a two-dimensional graph with the horizontal axis representing the time or order of sample batch collection and the vertical axis representing the test statistic values, $\bar{X}$ in this case. This chart has three horizontal lines:

Top line: The upper limit of the acceptance region,

$$\mu_0 + Z_{\alpha/2} \times \sigma_0/\sqrt{n} \qquad (8.1)$$

which is the upper control limit (UCL)

Bottom line: The lower limit of the acceptance region,

$$\mu_0 - Z_{\alpha/2} \times \sigma_0/\sqrt{n} \qquad (8.2)$$

which is the lower control limit (LCL)

Center Line: In-control value of the mean, μ_0

The chart is called an $\overline{X}$ chart. Dr. Walter Shewhart from Bell Labs introduced this and other control charts in 1923. He recommended that α be set equal to 0.0026. As $\alpha/2 = 0.0013$ and $Z_{\alpha/2} = 3.00$, the control limits become LCL $= \mu_0 - 3\sigma_0/\sqrt{n}$ and UCL $= \mu_0 + 3\sigma_0/\sqrt{n}$. Once the chart is set up, sample batches are collected every h time units, and $\overline{X}$ values are computed and plotted on the chart. The process is allowed to run if the plotted $\overline{X}$ values fall within the control limits. Whenever an $\overline{X}$ value falls outside the control limits (below the lower control limit or above the upper control limit), the process is stopped and diagnosed for assignable causes.

8.2.2 Probability of Type II Error (β)

$H_0: \mu = \mu_0$; $H_1: \mu \neq \mu_0$.

$$P[\text{Type II error}] = P[\text{Accept } H_0/\text{it is not true or } H_1 \text{ is true}]$$
$$= P[\text{Concluding that the process is in control when it is out of control}]$$
$$= P[\mu_0 - Z_{\alpha/2} \times \sigma_0/\sqrt{n} < \overline{X} < \mu_0 + Z_{\alpha/2} \times \sigma_0/\sqrt{n} | \mu \neq \mu_0]$$
$$(8.3)$$

Here, H_1 simply states that $\mu \neq \mu_0$, which is not sufficient to compute $P[\text{Type II error}]$. We need a specific value for μ that is not equal to μ_0. Let this be equal to μ_1 ($\mu_1 \neq \mu_0$):

$$P[\text{Type II error}] = P[\mu_0 - Z_{\alpha/2} \times \sigma_0/\sqrt{n} < \overline{X} < \mu_0 + Z_{\alpha/2} \times \sigma_0/\sqrt{n} | \mu = \mu_1]$$

As the probability distribution of $\overline{X}$ when H_1 is true (process is out of control) is approximately normal with a mean $= \mu_1$ and variance $= \sigma_0^2/n$,

$$P[\text{Type II error}] = P[(\mu_0 - Z_{\alpha/2} \times \sigma_0/\sqrt{n} - \mu_1)/\sigma_0/\sqrt{n} < Z$$
$$< (\mu_0 + Z_{\alpha/2} \times \sigma_0/\sqrt{n} - \mu_1)/\sigma_0/\sqrt{n}]$$
$$= P[(\mu_0 - \mu_1)\sqrt{n}/\sigma_0 - Z_{\alpha/2} < Z < (\mu_0 - \mu_1)\sqrt{n}/\sigma_0 + Z_{\alpha/2}]$$
$$(8.4)$$

The probabilities of Type I and II error are illustrated in Figure 8.1.

Because of β, the process is run in its out-of-control state for some time before an $\overline{X}$ falls outside the control limits. Defective parts will be manufactured during this interval, because μ is either $<\mu_0$ or $>\mu_0$. An important "measure" of interest in such cases is the average length of time during which the process is run, after it has gone out of control ($\mu \neq \mu_0$). This is called the *average run length* (ARL).

If α is small (0.0026 usually), β could be large, resulting in a large ARL and hence a large expected cost of rejection. Hence, other stopping rules besides an $\overline{X}$ falling outside the control limits are needed to stop the process.

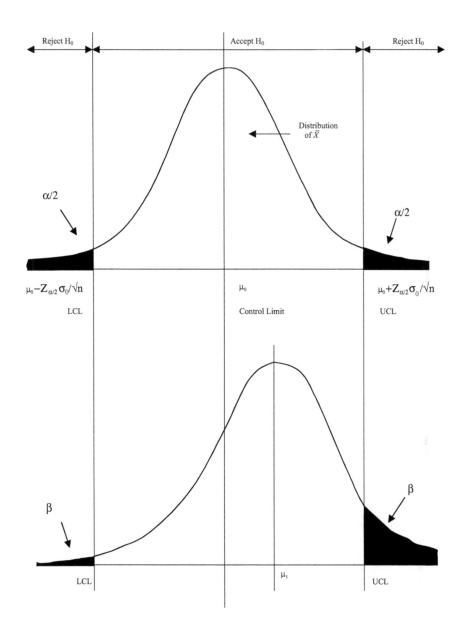

FIGURE 8.1
Probabilities of Type I and II errors.

Some rules recommended in QS-9000 are given below:

1. Seven points in a row on one side of the center line
2. Seven points in a row consistently going up or coming down
3. Substantially more than 2/3 of the points close to the center line
4. Substantially fewer than 2/3 of the points close to the center line

8.3   Variable Control Charts

Variable control charts are used when quality is measured as variables (length, weight, tensile strength, etc.). The main purpose of the variable control charts is to monitor the process mean and the standard deviation.

8.3.1   Monitoring Process Mean

Table 8.1 contains sample observations that will be used to illustrate calculation of control limits. The data set has 20 batches, each batch consisting of five observations ($n = 5$). The control limits are

$$
\begin{aligned}
\text{LCL} &= \mu_0 - 3\sigma_0/\sqrt{n} \\
\text{UCL} &= \mu_0 + 3\sigma_0/\sqrt{n}
\end{aligned}
\tag{8.5}
$$

When both μ_0 and σ_0 are known (specified and not estimated before running the process for regular production),

$$
\begin{aligned}
\text{LCL} &= \mu_0 - A\sigma_0 \\
\text{UCL} &= \mu_0 + A\sigma_0
\end{aligned}
\tag{8.6}
$$

TABLE 8.1

Data for $\bar{X}$, R, and s Charts[a]

Batch # (i)	Obs. 1	Obs. 2	Obs. 3	Obs. 4	Obs. 5	$\bar{X}$	R	s
1	4.5	4.6	4.5	4.4	4.4	4.48	0.2	0.084
2	4.6	4.5	4.4	4.3	4.1	4.38	0.5	0.192
3	4.6	4.1	4.4	4.4	4.1	4.32	0.5	0.217
4	4.4	4.3	4.4	4.2	4.3	4.32	0.2	0.084
5	4.3	4.3	4.4	4.2	4.3	4.30	0.2	0.071
6	4.6	4.6	4.2	4.5	4.5	4.46	0.4	0.167
7	4.1	4.3	4.6	4.5	4.2	4.34	0.5	0.207
8	4.5	4.5	4.4	4.6	4.4	4.48	0.2	0.084
9	4.4	4.2	4.6	4.6	4.2	4.40	0.4	0.200
10	4.2	4.2	4.2	4.5	4.2	4.26	0.3	0.134
11	4.3	4.2	4.3	4.4	4.2	4.28	0.2	0.084
12	4.4	4.4	4.4	4.4	4.1	4.34	0.3	0.134
13	4.3	4.2	4.4	4.6	4.6	4.42	0.4	0.179
14	4.2	4.4	4.4	4.1	4.4	4.30	0.3	0.141
15	4.2	4.3	4.1	4.5	4.6	4.34	0.5	0.207
16	4.6	4.4	4.3	4.5	4.1	4.38	0.5	0.192
17	4.6	4.6	4.6	4.2	4.5	4.50	0.4	0.173
18	4.4	4.6	4.3	4.1	4.3	4.34	0.5	0.182
19	4.3	4.6	4.2	4.2	4.1	4.28	0.5	0.192
20	4.2	4.5	4.1	4.4	4.4	4.32	0.4	0.164
Average						4.36	0.37	0.15

[a] $\mu_0 = 4.35$; $\sigma_0 = 0.1708$.

where

$$A = \frac{3}{\sqrt{n}} \tag{8.7}$$

Values of A are listed in Table A.4 in the Appendix.

When μ_0 and σ_0 are estimated, μ_0 is estimated by $\bar{X}$ and σ_0 is estimated by R/d_2 or s/c_4, where R is the sample range and s is the sample standard deviation.

8.3.1.1 μ_0 Estimated by $\bar{X}$ and σ_0 Estimated by R/d_2

Let $\bar{\bar{X}}$ be the grand average of more than one $\bar{X}$ and $\bar{R}$ be the grand average of more than one R. Then, μ_0 is estimated by $\bar{\bar{X}}$ and σ_0 is estimate by $\bar{R}/d_2$. In Table 8.1, $\bar{\bar{X}}$ is the grand average of 20 $\bar{X}$ values (each $\bar{X}$ is from a batch of 5 observations) and $\bar{R}$ is the average of 20 range (R) values (each R is from a batch of 5 observations). $\bar{\bar{X}} = 4.36$, $\bar{R} = 0.37$, and $n = 5$.

$$LCL = \mu_0 - 3\sigma_0/\sqrt{n} = \bar{\bar{X}} - 3\bar{R}/\sqrt{n}$$
$$= \bar{\bar{X}} - A_2\bar{R}, \quad \text{where } A_2 = 3/(d_2\sqrt{n}) \tag{8.8}$$

$$UCL = \bar{\bar{X}} + A_2\bar{R} \tag{8.9}$$

$$CL = \bar{\bar{X}} \tag{8.10}$$

Values of A_2 are tabulated in Table A.4.

Example 8.1
$\bar{\bar{X}} = 4.36$, $\bar{R} = 0.37$, $n = 5$, and A_2 from Table A.4 = 0.5768.

LCL = 4.36 − (0.5768 × 0.37) = 4.15
UCL = 4.36 + (0.5768 × 0.37) = 4.57
CL = 4.36

The 20 $\bar{X}$ values are plotted on the $\bar{X}$ chart in Figure 8.2.

8.3.1.2 μ_0 Estimated by $\bar{\bar{X}}$ and σ_0 Estimated by s/c_4

Let $\bar{s}$ be the grand average of more than one s (sample standard deviation). Then σ_0 is estimated by $\bar{s}/c_4$:

$$LCL = \bar{\bar{X}} - 3(\bar{s}/c_4)/\sqrt{n} = \bar{\bar{X}} - (3\bar{s})/(c_4\sqrt{n}) = \bar{\bar{X}} - A_3\bar{s} \tag{8.11}$$

where

$$A_3 = 3/(c_4\sqrt{n}) \tag{8.12}$$

$$UCL = \bar{\bar{X}} + A_3\bar{s} \tag{8.13}$$

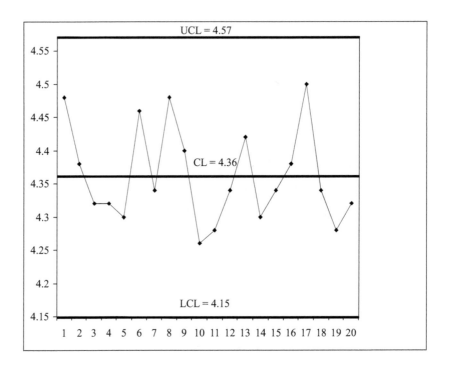

FIGURE 8.2
$\bar{X}$ chart.

Values of A_3 are tabulated in Table A.4 in the Appendix.

Example 8.2
$\bar{\bar{X}} = 4.36$, $\bar{s} = 0.154$, $n = 5$, and A_3 from Table A.4 = 1.4273.

    LCL = 4.36 − (1.4273 × 0.154) = 4.14
    UCL = 4.36 + (1.4273 × 0.154) = 4.58
    CL = 4.36

8.3.2 Monitoring Process Standard Deviation

The hypotheses tested are H_0: $\sigma = \sigma_0$; H_1: $\sigma \neq \sigma_0$. The general formula for any control chart is

Expected value of the test statistic $\pm Z_{\alpha/2}$

                      × standard deviation of the test statistic   (8.14)

Both the expected value and the standard deviation of the test statistic are obtained assuming that the process is in control. Also, it is assumed that the test statistic is approximately normally distributed. There are two test statistics (unbiased estimators) for testing these hypotheses: R/d_2 and s/c_4. Both are assumed to be approximately normally distributed.

8.3.2.1 Using R/d₂ as the Test Statistic

$$E[R/d_2] = \sigma_0; \quad \text{standard deviation } R/d_2 = (d_3/d_2) \times \sigma_0 \quad (8.15)$$

The control limits are

$$\sigma_0 \pm Z_{\alpha/2} \times (d_3/d_2) \times \sigma_0 = \sigma_0 [1 \pm 3 \times (d_3/d_2)] \quad (8.16)$$

if $\alpha = 0.0026$. The center line is located at σ_0. The values of the constant d_3 in Eqs. (8.15) and (8.16) are given in Table A.4 in the Appendix.

Values of R/d_2 are plotted on this control chart. Maintaining this chart can be simplified by using R as the test statistic instead of R/d_2. The control limits when values of R are plotted are $\sigma_0 (d_2 \pm 3 \times d_3)$ and the center line is at $\sigma_0 \times d_2$.

$$LCL = \sigma_0 (d_2 - 3 \times d_3) = D_1 \sigma_0 \quad (8.17)$$

and

$$UCL = \sigma_0 (d_2 + 3 \times d_3) = D_2 \sigma_0 \quad (8.18)$$

where

$$D_1 = (d_2 - 3 \times d_3) \quad (8.19)$$

and

$$D_2 = (d_2 + 3 \times d_3) \quad (8.20)$$

The values of D_1 and D_2 are given in Table A.4.

As $D_1 = (d_2 - 3 \times d_3)$ is negative when $n \leq 6$, D_1 is set to 0, as R is positive. The limits and the center line require knowledge of the in-control standard deviation, σ_0, which may be difficult in many real-life situations. In such cases, σ_0 has to be estimated. The unbiased estimator of σ_0 is R/d_2. Let us replace σ_0 in the above formulas for the control limits and the center line by R/d_2.

The control limits are

$$LCL = \sigma_0(d_2 - 3 \times d_3) = \frac{\overline{R}}{d_2} (d_2 - 3 \times d_3) = \overline{R}(1 - 3 \times d_3/d_2)$$

$$= D_3 \times \overline{R} \quad (8.21)$$

where

$$D_3 = 1 - 3 \times d_3/d_2 \tag{8.22}$$

$$\text{UCL} = \sigma_0(d_2 + 3 \times d_3) = \frac{\bar{R}}{d_2}(d_2 + 3 \times d_3) = \bar{R}(1 + 3 \times d_3/d_2)$$

$$= D_4 \times \bar{R} \tag{8.23}$$

where

$$D_4 = 1 + 3 \times d_3/d_2 \tag{8.24}$$

The values of D_3 and D_4 can be found in Table A.4 in the Appendix.

As $D_3 = (1 - 3 \times d_3/d_2)$ is negative when $n \le 6$, D_3 set equal to 0. The center line is at $\bar{R}$. This is called an R *chart*, because the values of the sample range, R, are plotted on this chart.

Example 8.3

From Table 8.1, $\bar{R} = 0.37$ and $n = 5$. The control limits are

$D_3 = 0$
$D_4 = 2.1144$
$\text{LCL} = 0$
$\text{UCL} = 0.37 \times 2.1144 = 0.78$

The center line is at $\bar{R} = 0.37$.

8.3.2.2 *Using s/c₄ as the Test Statistic*

$$E[s/c_4] = \sigma_0; \text{ standard deviation } s/c_4 = \frac{\sqrt{1 - c_4^2}}{c_4} \sigma_0 \tag{8.25}$$

The control limits are

$$\sigma_0 \pm 3 \times \frac{\sqrt{1 - c_4^2}}{c_4} \sigma_0 = \sigma_0 \left[1 \pm 3 \times \frac{\sqrt{1 - c_4^2}}{c_4} \right] \tag{8.26}$$

The center line is set at σ_0. Values of s/c_4 are plotted on this chart. Maintaining this chart can be simplified by using s as the test statistic instead of s/c_4. The control limits of a chart on which the values of s are plotted are $c_4 \times \sigma_0[1 \pm 3 \times \sqrt{1 - c_4^2}/c_4] = \sigma_0[c_4 \pm 3 \times \sqrt{1 - c_4^2}]$ and the center line is at $\sigma_0 c_4$.

$$\text{LCL} = \sigma_0[c_4 - 3 \times \sqrt{1 - c_4^2}] = B_5 \sigma_0 \tag{8.27}$$

and

$$\text{UCL} = \sigma_0[c_4 + 3 \times \sqrt{1 - c_4^2}] = B_6\sigma_0 \tag{8.28}$$

where

$$B_5 = c_4 - 3 \times \sqrt{1 - c_4^2} \tag{8.29}$$

and

$$B_6 = c_4 + 3 \times \sqrt{1 - c_4^2} \tag{8.30}$$

The values of B_5 and B_6 are given in Table A.4.

The limits and the center line require knowledge of the in-control standard deviation, σ_0, which may be difficult in many real-life situations. In such cases, σ_0 has to be estimated. The unbiased estimator of σ_0 is $\bar{s}/c_4$. Let us replace σ_0 in the above formulas for the control limits and the center line by $\bar{s}/c_4$.

The control limits are

$$\text{LCL} = \sigma_0[c_4 - 3 \times \sqrt{1 - c_4^2}] = \frac{\bar{s}}{c_4}[c_4 - 3 \times \sqrt{1 - c_4^2}]$$

$$= \bar{s}\left[1 - 3 \times \frac{\sqrt{1 - c_4^2}}{c_4}\right]$$

$$= B_3 \times \bar{s} \tag{8.31}$$

where

$$B_3 = 1 - 3 \times \frac{\sqrt{1 - c_4^2}}{c_4} \tag{8.32}$$

$$\text{UCL} = \sigma_0[c_4 + 3 \times \sqrt{1 - c_4^2}] = \bar{s}/c_4[c_4 + 3 \times \sqrt{1 - c_4^2}]$$

$$= \bar{s}\left[1 + 3 \times \frac{\sqrt{1 - c_4^2}}{c_4}\right]$$

$$= B_4 \times \bar{s} \tag{8.33}$$

where

$$B_4 = 1 + 3 \times \frac{\sqrt{1 - c_4^2}}{c_4} \qquad (8.34)$$

The values of B_3 and B_4 can be found in Table A.4.

As $B_3 = (1 - 3 \times \sqrt{(1 - c_4^2)}/c_4)$ is negative when $n \leq 5$, it is set equal to 0. This is called an s chart, because the values of the sample standard deviation, s, are plotted on this chart.

Example 8.4

Please refer to Table 8.1. From the table, $\bar{s} = 0.15$ and $n = 5$. The control limits are

$B_3 = 0$

$B_4 = 2.0889$

$LCL = 0$

$UCL = 2.0889 \times 0.154 = 0.322$

The center line is at $\bar{s} = 0.154$.

8.3.2.3   Summary

While monitoring the mean and standard deviation of a process, two control charts are required:

1. $\bar{X}$ chart (for testing μ) and R chart (for testing σ)
2. $\bar{X}$ chart (for testing μ) and s chart (for testing σ)

8.3.2.4   Subgroup or Batch Size and Frequency

Based on the recommendation of QS-9000, the sample batches should be chosen in such a manner that the opportunities for variation among the units within a batch are small. The recommended subgroup size (n) for an initial study is 4 to 5. It should be kept in mind that the components within a subgroup are produced under very similar conditions and that no systematic variation is present among the units within the same subgroup. The variation present among the units of a subgroup should be due to the natural inherent variation present in the process only. This is called the *common cause variation*. The control charts compare this variation with the variation between subgroups, which is due to the "special" or "assignable" causes. As the purpose of the charts is to detect changes in the process over time, batches should be collected often enough to do so. The interval between successive batches could be shorter initially. As the process becomes stable, this interval could be increased. For estimation of the in-control mean and standard deviation, it is recommended that 25 or more subgroups containing about 100 or more components are collected. In the Design of Control Charts section (8.5), the sample size (n) and the interval between successive batches (h) will be considered as decision variables to minimize the total expected cost per unit time.

8.3.3   Some Special Charts

There are situations where it may be difficult to take a sample of size greater than one or when only one measurement is meaningful each time. Some examples of these situations are

1. The production rate is very slow or the batch size or the lot size is very small.
2. In continuous processes (such as chemical processes), measurements on some quality characteristics, such as the viscosity of a paint or the thickness of insulation on a cable, will vary only a little between successive observations.

In these situations, it is not possible to use the $\overline{X}$ and R or s charts we studied earlier. The following are some of the charts that can be used in such cases.

8.3.3.1   *Individual Measurement Chart or X Chart or Run Chart and Moving Range Chart*

Both the in-control mean and in-control standard deviation are estimated.

Example 8.5

The ten observations given in Table 8.2 are the densities of a compound collected at intervals of 15 minutes each from a chemical process. These observations will be used to compute the limits of the X chart and the moving range chart.

8.3.3.1.1   *X Chart*

The control limits of an $\overline{X}$ chart are $\mu_0 \pm 3\sigma_0/\sqrt{n}$ (assuming $\alpha = 0.0026$). In the X chart, individual measurements are plotted, hence $n = 1$. So the control limits of an X chart are $\mu_0 \pm 3\sigma_0$. As μ_0 is estimated by $\overline{X}$ and σ_0 by $\overline{R}/d_2$, the limits become:

$$\overline{X} \pm 3(\overline{R}/d_2) \tag{8.35}$$

TABLE 8.2

Data for X Chart and Moving Range Chart

Observation Number	Observation	Moving Range
1	10.42	—
2	10.89	0.47
3	9.86	1.03
4	9.97	0.11
5	10.52	0.55
6	10.21	0.31
7	10.99	0.78
8	11.23	0.24
9	11.21	0.02
10	11.14	0.07
Average	10.64	0.398

As the moving range is calculated from two successive observations most of the time, and $d_2 = 1.128$ when $n = 2$, the control limits are

$$\bar{X} \pm 2.66\bar{R} \tag{8.36}$$

In the example, $\bar{X} = 10.64$ and $\bar{R} = 0.398$, hence the limits are (9.58, 11.70). None of the ten X values are outside these limits; therefore, these can be used as the limits of the X chart.

8.3.3.1.2 Moving Range Chart

The control limits of the R chart are $D_3\bar{R}$ and $D_4\bar{R}$ (assuming $\alpha = 0.0026$). As the moving range is computed from two successive observations most of the time, D_3 and D_4 are 0 and 3.2672, respectively. Hence, the limits of the moving range chart are

$$(0, \ 3.2672\bar{R}) \tag{8.37}$$

In our example, the limits are (0, 1.30). As none of the nine range values is outside these limits, these limits can be used as the limits of the moving average chart.

8.3.3.2 Pre-Control Chart

Individual measurements are plotted on this chart. It is suitable for situations where the batch size is small. The control limits are based on the specification limits. As an example, let the specification limits for a quality characteristic be 0.5 ± 0.002 (LSL = 0.498; USL = 0.502). The center line of the chart is located at the nominal size (0.500). Horizontal lines are drawn at the upper specification limit (0.502) and the lower specification limit (0.498). In addition, horizontal lines are also drawn at nominal size $\pm 1/4 \times$ (USL – LSL).

In our example, these lines are at $0.5 - 1/4 \times (0.502 - 0.498) = 0.499$ and $0.5 + 1/4 \times (0.502 - 0.498) = 0.501$. The regions above the USL and below the LSL are called the *red zone*; the interval between (nominal size $- 1/4 \times$ total tolerance) and (nominal size $+ 1/4 \times$ total tolerance) is called the *green zone*; and the regions between the red and green zones are called the *yellow zone*.

The following rules are used while setting up the process:

1. Collect components and measure and plot the individual measurements on the chart until five consecutive values fall in the green zone.

2. If a value falls in the yellow zone, restart the count to obtain five consecutive pieces in the green zone. Do not adjust the process.

3. If two consecutive values fall in the yellow zone or one value falls in the red zone, adjust the process.

4. When five consecutive values fall in the green zone, approve the setup (the process is in-control) and start regular manufacture.

During regular manufacture, sample two consecutive components every h (say, 20) minutes and follow these rules:

1. If the first value falls in the green zone, do not plot the second value; continue the process.

2. If the first value falls in the red zone, stop the process and investigate.

3. If the first value falls in the yellow zone, then plot the second value. If it falls in the green zone, continue the process; otherwise, stop the process and investigate.

This chart is simple to maintain, which is very important. One main disadvantage is that the information presented by the chart regarding the variability of the process is incomplete.

8.3.3.3 D-NOM Charts

In these charts, the deviations of the characteristics from their respective nominal values are used as the observations. The calculations of the control limits are done in the same manner as in the regular $\bar{X}$ and R charts.

Example 8.6

Let us assume that two types of shafts, A and B, are produced on a machine. The nominal values of the quality characteristics of A and B are 30.00 and 20.00, respectively. Table 8.3 contains three sample batches of A and five batches of B. The sample size is three for all the batches.

TABLE 8.3

Data for D-NOM Chart

Batch #	Part Type	Obs. 1	Obs. 2	Obs. 3	Deviation of Obs. 1 from Its Nominal Value (x_1)	Deviation of Obs. 2 from Its Nominal Value (x_2)	Deviation of Obs. 3 from Its Nominal Value (x_3)	Test Statistic ($\bar{x}$)	Test Statistic (R)
1	A	30	31	32	0	1	2	1.0	2
2	A	29	30	31	−1	0	1	0	2
3	A	28	29	32	−2	−1	2	−0.33	4
4	B	20	22	21	0	2	1	0.67	3
5	B	20	22	19	0	2	−1	0.33	2
6	B	22	21	18	2	1	−2	0.33	4
7	B	20	19	18	0	−1	−2	−1.00	2
8	B	19	20	20	−1	0	0	−0.33	1

The control limits are

1. $\overline{X}$ chart
   LCL: $\overline{\overline{X}} - A_2 \, (n = 3) \, \overline{R} = 0.084 - 1.0231 \times 2.5 = -2.47$
   UCL: $\overline{\overline{X}} + A_2 \, (n = 3) \, \overline{R} = 0.084 + 1.0231 \times 2.5 = 2.64$
   CL: 0.084
2. R chart
   LCL: $D_3\overline{R} = 0$
   UCL: $D_4\overline{R} = 2.5743 \times 2.5 = 6.438$
   CL: $\overline{R} = 2.5$

The assumptions are

1. Process standard deviation is the same for all parts.
2. Sample size is constant.

8.3.3.4   Standardized $\overline{X}$ and R Charts

These are used, if the assumption that the standard deviation is the same cannot be satisfied. For the part type j test statistic, let X_{0j} be the target value for part type j and $\overline{R}_j$ be the average range of part type j. Then,

$$\overline{X} \text{ chart test statistic} = (\overline{X}_j - X_{0j})/\overline{R}_j \qquad (8.38)$$

where $(\overline{X}_j - X_{0j})$ is equal to $\Sigma_{i=1}^{n}(X_{ij} - X_{0j})/n$ and n is the equal sample size. The control limits are

LCL $= -A_2$
UCL $= +A_2$
CL $= 0.0$

The R chart test statistic $= R/\overline{R}_j$, and

LCL $= D_3$
UCL $= D_4$
CL $= 1.0$

Example 8.7

The test statistics are computed using the same data given in Table 8.3. The results are given in Table 8.4. For the $\overline{X}$ chart,

LCL $= -A_2 \, (n = 3) = -1.023$
UCL $= A_2 \, (n = 3) = 1.023$
CL $= 0.0$

TABLE 8.4

Data for the Standardized $\bar{X}$ and R Charts

Batch #	Part Type	Obs. 1	Obs. 2	Obs. 3	$\bar{x}$	R	$\bar{x}$ Chart Test Statistic	R Chart Test Statistic
1	A	0	1	2	1.00	2	$1/2.67 = 0.375$	$2/2.67 = 0.75$
2	A	−1	0	1	0.00	2	0	$2/2.67 = 0.75$
3	A	−2	−1	2	−0.33	4	−0.124	1.50
						$8/3 = 2.67$		
4	B	−1	2	1	0.67	2	$0.67/2.2 = 0.305$	$2/2.2 = 0.909$
5	B	0	2	−1	0.33	2	0.15	0.909
6	B	2	1	−2	0.33	4	0.15	1.818
7	B	0	−1	−2	−1.00	2	−0.455	0.909
8	B	−1	0	0	−0.33	1	−0.15	0.455
						$11/5 = 2.2$		

For the R chart,

$$LCL = D_3 = 0$$
$$UCL = D_4 \, (n = 3) = 2.575$$
$$CL = 1.00$$

There must be some logic for pooling parts. The target value X_{0j} and the average range R_j can be determined from past data.

8.3.3.5   Exponentially Weighted Moving-Average (EWMA) Control Chart

This chart is mainly used with individual observations; it was introduced by Roberts.[9] Let us assume that i observations have been collected from the process until now. The test statistic associated with the ith observation, x_i, is the exponentially weighted moving average associated with the ith observation:

$$z_i = \lambda x_i + (1 - \lambda)Z_{i-1} \tag{8.39}$$

where

$$Z_{i-1} = \lambda x_{i-1} + (1 - \lambda)Z_{i-2}$$

$$Z_{i-2} = \lambda x_{i-2} + (1 - \lambda)Z_{i-3}$$

and so on where $0 < \lambda \le 1$.
  When $i = 1$,

$$z_1 = \lambda x_1 + (1 - \lambda)z_0$$

where $z_0 = \mu_0$ (in-control mean or an estimate of the in-control mean).

Now, z_i can be written as:

$$
\begin{aligned}
z_i &= \lambda x_i + (1-\lambda)[\lambda x_{i-1} + (1-\lambda)Z_{i-2}] \\
&= \lambda x_i + \lambda(1-\lambda)x_{i-1} + (1-\lambda)^2 Z_{i-2} \\
&= \lambda x_i + \lambda(1-\lambda)x_{i-1} + (1-\lambda)^2[\lambda x_{i-2} + (1-\lambda)Z_{i-3}] \\
&= \lambda x_i + \lambda(1-\lambda)x_{i-1} + \lambda(1-\lambda)^2 x_{i-2} + (1-\lambda)^3 Z_{i-3} \\
&= \lambda x_i + \lambda(1-\lambda)x_{i-1} + \lambda(1-\lambda)^2 x_{i-2} + \lambda(1-\lambda)^3 x_{i-3} \\
&\quad + \cdots + \lambda(1-\lambda)^{i-j}x_j \cdots + \lambda(1-\lambda)^{i-1}x_1 + (1-\lambda)^i Z_0 \\
&= (1-\lambda)^i Z_0 + \lambda(1-\lambda)^{i-1}x_1 + \lambda(1-\lambda)^{i-2}x_2 + \cdots + \lambda(1-\lambda)^{i-j}x_j \\
&\quad + \cdots + \lambda(1-\lambda)^2 x_{i-2} + \lambda(1-\lambda)x_{i-1} + \lambda x_i \\
&= (1-\lambda)^i Z_0 + w_1 x_1 + w_2 x_2 + \cdots + w_j x_j + \cdots + w_{i-1}x_{i-1} + w_i x_i \quad (8.40)
\end{aligned}
$$

It can be seen from the expression of z_i that the weight given to the jth observation, x_j, is

$$
w_j = \lambda(1-\lambda)^{i-j} \quad\quad (8.41)
$$

Since $1 - \lambda \le 1$, $w_j = \lambda(1-\lambda)^{i-j}$ increases as j increases. That is, the weight assigned to the most recent observation is larger than the weights assigned to the previous observations.

Example 8.8

Let $\lambda = 0.8$, $i = 5$, and $\mu_0 = 1.0$. Then, the weights are $w_1 = 0.6(1-0.6)^{5-1} = 0.01536$; $w_2 = 0.6(1-0.6)^{5-2} = 0.0384$; $w_3 = 0.6(1-0.6)^{5-3} = 0.096$; $w_4 = 0.6(1-0.6)^{5-4} = 0.240$; and $w_5 = 0.6(0.4)^{5-5} = 0.6$. Also, $z_5 = 0.6x_5 + 0.24x_4 + 0.096x_3 + 0.0384x_2 + 0.01536x_1$. The test statistics if $x_1 = 0.98$, $x_2 = 0.97$, $x_3 = 1.00$, $x_4 = 1.02$, and $x_5 = 1.03$ (and $\mu_0 = 1.0$) are given in Table 8.5.

TABLE 8.5

Test Statistics for the EWMA Chart

Obs. # (i)	Obs. (x_i)	Test Statistic $z_i = \lambda x_i + (1-\lambda)z_{i-1}$
1	0.98	$z_1 = 0.6 \times 0.98 + (1-0.6) \times 1.0 = 0.988$
2	0.97	$z_2 = 0.6 \times 0.97 + (1-0.6) \times 0.988 = 0.977$
3	1.00	$z_3 = 0.6 \times 1.00 + (1-0.6) \times 0.977 = 0.991$
4	1.02	$z_4 = 0.6 \times 1.02 + (1-0.6) \times 0.991 = 1.008$
5	1.03	$z_5 = 0.6 \times 1.03 + (0.4) \times 1.008 = 1.021$

TABLE 8.6

Control Limits for EWMA Chart

Obs. # (i)	Observation (x_i)	Test Statistic (z_i)	LCL	UCL
1	0.98	0.988	$1 - 3.054 \times 0.0133$ $\times \sqrt{\{0.6/(2-0.6)[1-(1-0.6)^2]\}}$ $= 0.976$	$1 + 3.054 \times 0.0133$ $\times \sqrt{\{0.6/(2-0.6)[1-(1-0.6)^2]\}}$ $= 1.024$
2	0.97	0.977	0.974	1.026
3	1.00	0.991	0.973	1.027
4	1.02	1.008	0.973	1.027
5	1.03	1.021	0.973	1.027

The control limits are

$$\text{LCL} = \mu_0 - L\sigma\sqrt{\frac{\lambda}{(2-\lambda)}[1-(1-\lambda)^{2i}]} \tag{8.42}$$

$$\text{UCL} = \mu_0 + L\sigma\sqrt{\frac{\lambda}{(2-\lambda)}[1-(1-\lambda)^{2i}]} \tag{8.43}$$

and $\text{CL} = \mu_0$.

In the above expressions for LCL and UCL, σ_0 is the in-control standard deviation, λ is the parameter used in computing the test statistic, and L is the width of the control limits (which is $Z_{\alpha/2}$). Table 8.6 contains the control limits for the data in Example 8.8.

The parameters λ and L are selected to yield a specified average run length (ARL), which is the average length of time during which the process is run in its out-of-control state before it is stopped. Lucas and Saccucci[5] provide guidelines for selecting λ and L. In general, values of λ in the interval $0.05 \leq \lambda \leq 0.25$ and $L = 3$ work well in practice.[7] Smaller values of λ should be used to detect smaller shifts.

8.4   Attribute Control Charts

In cases where quality is measured as attributes (number of defects in a component or a product or a batch of components or products, number or proportion of defectives in a batch, etc.), attribute control charts are used.

8.4.1   Monitoring Proportion of Defectives in a Lot

The proportion of defectives in a lot is denoted by p. The hypotheses tested are $H_0: p = p_0$; $H_1: p \neq p_0$ (p_0 is some target value which should be as small as possible).

The test statistic is the sample proportion of defectives $\bar{p}$ or $\hat{p} = x/n$, where n = sample batch size and x = number of defectives in the sample batch. The control limit general formula is $E(\text{T.S.}) \pm Z_{\alpha/2}\,\text{S.D.}(\text{T.S.})$:

$$E(\bar{p}) = p_0; \quad \text{S.D.}(\bar{p}) = \sqrt{\frac{p_0(1-p_0)}{n}} \tag{8.44}$$

Hence, the control limits are

$$p_0 \pm Z_{\alpha/2}\sqrt{\frac{p_0(1-p_0)}{n}} \tag{8.45}$$

$$= p_0 \pm 3\sqrt{\frac{p_0(1-p_0)}{n}}, \quad \text{if } \alpha = 0.0026 \tag{8.46}$$

The center line is at p_0. If p_0 cannot be specified, it must be estimated from the data. Let us assume that m sample batches, each of size n, are collected and that the total number of defectives in these m sample batches is d. Then the estimate of the proportion of defectives per sample batch (of size n) is $\bar{\bar{p}} = d/(m \times n)$, and $\hat{p}_0 = \bar{\bar{p}}$.
Then the control limits are

$$\bar{\bar{p}} \pm 3\sqrt{\frac{\bar{\bar{p}}(1-\bar{\bar{p}})}{n}} \tag{8.47}$$

The center line is at $\bar{\bar{p}}$, and this is called a p chart.

Example 8.9
Let us assume that the readings in Table 8.7 are the number of defective items in 18 sample batches, each containing a total of 50 items ($n = 50$). The total number of defectives in the 18 sample batches collected is 229. As each sample batch has 50 items, the average fraction of defectives in these 18 sample batches is $\bar{\bar{p}} = 229/(18 \times 50) = 0.254$, and the limits using this as the estimate of p_0 are

$$\text{LCL} = 0.254 - 3\sqrt{0.254(1-0.254)}/50 = 0.069$$
$$\text{UCL} = 0.254 + 3\sqrt{0.254(1-0.254)}/50 = 0.439$$

The center line is located at 0.254.
In this example, all the $\bar{p}$ values are within the control limits computed. Hence, these limits can be used for monitoring p. If one or more $\bar{p}$ values falls outside the limits, then these values have to be removed and a new $\bar{\bar{p}}$ is

TABLE 8.7

Data for p Chart

Batch Number (i)	Number of Defectives (x)	Proportion of Defectives ($\bar{p}$)
1	9	0.18
2	10	0.20
3	11	0.22
4	13	0.26
5	13	0.26
6	8	0.16
7	18	0.36
8	12	0.24
9	11	0.22
10	8	0.16
11	14	0.28
12	21	0.42
13	18	0.36
14	10	0.20
15	8	16
16	18	0.36
17	19	0.38
18	8	0.16

computed. This procedure has to be repeated until all the $\bar{p}$ values used in the estimation of $\bar{\bar{p}}$ are within the control limits.

8.4.2 Monitoring Number of Defectives in a Lot

The number of defectives in a lot is denoted by np. The hypotheses tested are H_0: mean number of defectives $= np_0$; H_1: mean number of defectives $\neq np_0$.

The test statistic used is the number of defectives (x) in a sample batch of size n, which is denoted as np. The limits of this chart are obtained by multiplying the control limits of the p chart by n:

$$n \times \left[p_0 \pm 3 \sqrt{\frac{p_0(1-p_0)}{n}} \right], \quad \text{if } \alpha = 0.0026$$

$$= np_0 \pm 3\sqrt{np_0(1-p_0)} \tag{8.48}$$

In Eq. (8.48), p_0 can be replaced by its estimate $\bar{\bar{p}}$. The center line is located at np_0, and this is called an np chart.

Example 8.10

Using the data from Table 8.7, the control limits are computed as follows:

$\bar{\bar{p}} = 0.254$

$n = 50$

$$\text{LCL} = 50 \times 0.254 - 3\sqrt{50 \times 0.254 \times (1 - 0.254)} = 3.45.$$
$$\text{UCL} = 50 \times 0.254 + 3\sqrt{50 \times 0.254 \times (1 - 0.254)} = 21.95.$$

The center line is located at $50 \times 0.254 = 12.70$.

8.4.3 Monitoring Number of Defects

If the quality of a component or product is measured in terms of the number of defects per component or product or batch, then a c chart is used. The letter c here denotes the number of defects per component or product (or per some appropriate unit of the product—for example, number of defects per 10 yards of a cable or per 1 yd^2 of an aluminum sheet, etc.) or the number of defects in a sample of size n. The hypotheses tested here are

H_0: mean number of defects per piece or some unit or per batch $= c_0$.

H_1: mean number of defects per piece or some unit or product $\neq c_0$.

The test statistic is the number of defects per the appropriate unit, which is c. The expected value and the standard deviation are (assuming that c obeys a Poisson distribution) $E(c) = c_0$; S.D.$(c) = c_0$.
Hence, the control limits are $c_0 \pm Z_{\alpha/2}\sqrt{c_0}$, which becomes:

$$c_0 \pm 3\sqrt{c_0}, \quad \text{if } \alpha = 0.0026 \tag{8.49}$$

The center line is at c_0. If c_0 cannot be specified, it can be estimated by the mean number of defects from one or more sample batches, which is denoted by $\bar{c}$. Then, the control limits become $\bar{c} \pm 3\sqrt{c}$, if $\alpha = 0.0026$. The center line is located at $\bar{c}$.

Example 8.11
The observations in Table 8.8 are the number of defects in five sample batches, each containing 15 items. Assuming that the process was in control

TABLE 8.8

Data for c Chart

Batch # (i)	Number of Defects in Batch (c)
1	18
2	12
3	7
4	9
5	16

when these observations were collected, compute the limits of the c chart:

$$\bar{c} = (18 + 12 + 7 + 9 + 16)/5 = 12.4$$

The limits are LCL $= 12.4 - 3\sqrt{12.4} = 1.84$ and UCL $= 12.4 + 3\sqrt{12.4} = 22.96$. The center line is located at 12.4. All the c values used in the estimation of $\bar{c}$ are within the control limits.

8.4.4 Monitoring Average Number of Defects

If the quality is measured in terms of the *average* number of defects per unit (and not the *actual* number of defects per unit or the actual number of defects per sample batch), then a U chart is used. Here, U denotes the average number of defects per unit and c denotes the actual number of defects per sample batch containing n items.

If U_0 denotes the in-control or target mean of the average defects per unit, then the hypotheses being tested are

H_0: mean number of average defects per unit $= U_0$.

H_1: mean number of average defects per unit $\neq U_0$.

The test statistic used is the average number of defects per unit, or U. The expected value and standard deviation of U are

$$E(U) = U_0$$
$$\text{Standard deviation } (U) = \sqrt{(U_0/n)}$$
$$U = c/n$$
$$E(C) = c_0$$
$$\text{Var } (C) = c_0 \qquad\qquad (8.50)$$
$$E(U) = c_0/n = U_0$$
$$\text{Var } (U) = \text{Var } (C)/n^2 = c_0/n^2 = nU_0/n^2 = U_0/n$$

where n is the sample size. Hence, the control limits of the U chart are

$$U_0 \pm Z_{\alpha/2}\sqrt{(U_0/n)}$$

which becomes:

$$U_0 \pm 3\sqrt{(U_0/n)}, \quad \text{when } \alpha = 0.0026 \qquad\qquad (8.51)$$

The center line is at U_0. If U_0 cannot be specified, then it is estimated by the sample mean number of defects per unit from observations collected when

TABLE 8.9

Data for U Chart (I)

Batch # (*i*)	Number of Defects in Batch (*c*)	Average Number of Defects per Unit (*U*)
1	18	$18/15 = 1.2$
2	12	$12/15 = 0.8$
3	7	$7/15 = 0.47$
4	9	$9/15 = 0.6$
5	16	$16/15 = 1.07$

the process is in control. Let this estimate be $\bar{\bar{U}}$. Then the control limits are

$$\bar{\bar{U}} \pm 3\sqrt{(\bar{\bar{U}}/n)} \tag{8.52}$$

The center line is located at $\bar{\bar{U}}$.

Example 8.12

The data set in Table 8.9 contains the number of defects in five sample batches, each containing 15 items. Assuming that the process was in control when these observations were collected, compute the control limits of the U chart.

There are two options for calculating the average number of defects per unit (U). Both give the same value, as the sample sizes are equal:

$\bar{\bar{U}} = [1.2 + 0.8 + 0.47 + 0.6 + 1.07]/5 = 0.83$, or

$\bar{\bar{U}} = [18 + 12 + 7 + 9 + 16]/(5 \times 15) = 0.83$

In this case,

$\text{LCL} = 0.83 - 3\sqrt{(0.83/15)} = 0.12.$

$\text{UCL} = 0.83 + 3\sqrt{(0.83/15)} = 1.54.$

The center line is located at 0.83.

Example 8.13

The data on the number of defects found in five sample batches with unequal sample batch sizes are given in Table 8.10. Compute the control limits of the U chart.

In this example, the sample sizes are not equal, hence each sample batch will have its own control limits. This is because the control limits, $\bar{\bar{U}} \pm$

TABLE 8.10

Data for U Chart (II)

Batch # (i)	Sample Size (n_i)	Number of Defects in Batch (c_i)	Average Number of Defects per Unit (U_i)
1	2	5	$5/2 = 2.5$
2	3	6	$6/3 = 2.0$
3	2	2	$2/2 = 1.0$
4	5	7	$7/5 = 1.4$
5	8	9	$9/8 = 1.125$
Total	20	29	—

TABLE 8.11

Control Limits of U Chart

Batch # (i) (n_i)	U_i	LCL_i	UCL_i
1 (2)	$5/2 = 2.5$	0	4.30
2 (3)	$6/3 = 2.0$	0	3.81
3 (2)	$2/2 = 1.0$	0	4.30
4 (5)	$7/5 = 1.4$	0	3.31
5 (8)	$9/8 = 1.125$	0.26	2.96

$3\sqrt{(\overline{U}/n_i)}$, are functions of the sample size. However, $\overline{U}$ is computed using all the observations:

$$\overline{U} = [2.5 + 2.0 + 1.0 + 1.4 + 1.125]/5 = 1.61, \text{ or}$$
$$\overline{U} = [5 + 6 + 2 + 7 + 9]/[2 + 3 + 2 + 5 + 8] = 29/20 = 1.45$$

Either can be used to compute the control limits. We will use 1.61, hence the control limits are $1.61 \pm 3\sqrt{(1.61/n_i)}$, where the subscript i denotes batch i.
When $i = 1$:

$$LCL = 1.61 - 3\sqrt{(1.61/2)} = -1.08 \text{ which is set to 0.}$$
$$UCL = 1.61 + 3\sqrt{(1.61/2)} = 4.30.$$

When $i = 2$:

$$LCL = 1.61 - 3\sqrt{(1.61/3)} = -0.59 \text{ which is set to 0.}$$
$$UCL = 1.61 + 3\sqrt{(1.61/3)} = 3.81.$$

The control limits of the five batches are given in Table 8.11. It can be seen that all the U_i are within the respective control limits.

8.5 Design of Control Charts

In this section, the design of an $\overline{X}$ chart will be described, based on the work by Duncan.[3] The procedure used in the design of an $\overline{X}$ chart can be used in the design of other control charts, with suitable modifications. From Eqs. (8.1) and (8.2), the control limits of an $\overline{X}$ chart are $(\mu_0 \pm Z_{\alpha/2} \sigma_0/\sqrt{n})$, from which it can be seen that the variables required to construct an $\overline{X}$ chart are

1. μ_0 – In-control value of the process mean (specified or estimated)
2. α – Probability of Type I error (commonly used value is 0.0026)
3. σ_0 – In-control value of the process standard deviation (specified or estimated)
4. n – Sample size
5. h – Length of interval between successive sample batches

In the design of an $\overline{X}$ chart by Duncan,[3] n, h, and α are the decision variables. The following assumptions are made:

1. The process always starts in its in-control state.
2. When the process is in the in-control state, the mean (μ) of the quality characteristic is μ_0 (unknown).
3. A single assignable cause occurs at random (for example, a tool breaking) and results in a shift of magnitude δ in the mean from μ_0 to either ($\mu_0 + \delta$) or ($\mu_0 - \delta$). The out-of-control mean is denoted by μ_1.
4. If the process starts in the in-control state, the time interval during which the process remains in that in-control state is exponentially distributed with a mean $1/\lambda$ (known).
5. The process is monitored by an $\overline{X}$ chart with the center line at μ_0, UCL at $\mu_0 + k\sigma_0/\sqrt{n}$, and LCL at $\mu_0 - k\sigma_0/\sqrt{n}$, where $k = Z_{\alpha/2}$ (σ_0 is known; k and n are decision variables).
6. Sample batches of size n are taken at intervals of h time units (n and h are decision variables) and $\overline{X}$ is plotted on the chart.
7. When any point falls outside the control limits, a search for the assignable cause is initiated. The time required to find the assignable cause is a constant equal to D.
8. Once the cause is located, it is removed and the process is restarted in its in-control state (with $\mu = \mu_0$).

The objective is to minimize the total expected costs per time, including: sampling cost, cost of running the process in an out-of-control state, cost of

stopping the process unnecessarily (cost of false alarm), cost of investiga-
tions, and cost of rectification. The unit costs that Duncan used will be introduced as and when required. It is an unconstrained optimization problem
with three decision variables (n, h, and k), out of which two are continuous
variables (h and k) and one (n) is an integer variable.

Because of the assumptions made, the process goes through a cycle consisting of the following:

1. In-control period
2. Out-of-control period until detection
3. Time to take a sample batch and interpret the results
4. Time to find the assignable cause and fix it

The total length of the cycle is a random variable because intervals 1 and 2
are random variables. The cycle renews itself probabilistically at every start
and the lengths of the cycles are independent and identically distributed random variables. Hence, this cycle is a renewal cycle and this stochastic process
is a renewal process.[10] Our objective in this problem is to minimize the total
expected cost per unit time, denoted by $E(TC)$. As per the Renewal Reward
Theorem, the total expected cost per unit time is equal to

$$E(TC) = E(C)/E(CT) \qquad (8.53)$$

where $E(C)$ is the expected cost/cycle and $E(CT)$ is the expected length of a
cycle. Now the problem reduces to that of finding $E(C)$ and $E(CT)$. Let us first
obtain the expected length of a cycle, $E(CT)$. We will find the expected length
of each of the nonoverlapping four segments in the cycle and add them to
find the expected length of the whole cycle.

8.5.1 In-Control Period

This is the part of the cycle during which the process stays in control. In other
words, during this entire period the process mean stays at the in-control
value, μ_0. As per assumption 4, this period is an exponentially distributed
random variable with a mean $1/\lambda$. Hence,

$$E \text{ [In-control period]} = 1/\lambda \qquad (8.54)$$

8.5.2 Out-of-Control Period

This period starts when the process goes out of control (that is, the process
mean μ has changed from the target value μ_0 to μ_1) and ends when the user
detects it because of an $\overline{X}$ falling outside the control limits on the control
chart. If we can obtain the expected number of trials required for the

detection of the out-of-control state of the process (that is, the expected number of $\overline{X}$ values to be plotted before the first $\overline{X}$ falls outside the control limits), then the expected length of the period until detection, which we will denote by $E(B)$, can be obtained by multiplying this expected number of $\overline{X}$ values and the interval length, which is h. The number of $\overline{X}$ values to be plotted before the first $\overline{X}$ falls outside the control limits follows a geometric distribution with a mean of $1/(1 - \beta)$, where β is the probability of Type II error defined in Eq. (8.4). Hence, the expected length of the period until detection is

$$E(B) = h/(1 - \beta) \tag{8.55}$$

There is an overlap between the in-control period and $E(B)$. Let us denote this overlap by L. The expected value of this period—that is, $E(L)$—is to be subtracted from $E(B)$ in order to obtain the expected length of the out-of-control period until detection.

The problem in finding $E(L)$ is that we do not know at what interval the process goes out of control. We can solve this problem by assuming that the out-of-control state starts in the ith interval $[(i - 1)h, ih]$ and conditioning on that event.

Let $A_i \equiv$ out of control that occurs during the ith interval. That is,

$$A_i = [(i - 1)h \leq S < ih]$$

where S is the time during which the process stays in control. Hence,

$$P(A_i) = P[(i - 1)h < S < ih] \tag{8.56}$$

The expected length of L, given that failure occurs (out-of-control state starts) during the ith interval, is given by:

$$E[L/A_i] = E[S/A_i] - (i - 1)h \tag{8.57}$$

Once the $E[L/A_i]$ for all i ($i = 1, 2, \ldots$), are obtained, $E(L)$ can be found as:

$$E(L) = \sum_{i=1}^{\infty} E[L/A_i]P[A_i] \tag{8.58}$$

Let $f(s)$ be the probability density function of S, which is the time during which the process is in control. From the definition of conditional expectation,

$$E[S/A_i] = \int_{(i-1)h}^{ih} S f(s/A_i)\, ds \tag{8.59}$$

The conditional probability density function, $f(s/A_i)$ is

$$f(s/A_i) = f(s)/P[A_i] = \lambda e^{-\lambda s}/P[A_i] \tag{8.60}$$

where

$$P[A_i] = \int_{(i-1)h}^{ih} f(s)\,ds$$
$$= e^{-(i-1)h\lambda} - e^{-\lambda ih} \tag{8.61}$$

Now, Eq. (8.60) can be written as:

$$f(s/A_i) = \lambda e^{-\lambda s}/[e^{-(i-1)h\lambda} - e^{-\lambda ih}], \quad (i-1)h < s < ih \tag{8.62}$$

Hence, Eq. (8.59) is

$$E[S/A_i] = \int_{(i-1)h}^{ih} s\lambda e^{-\lambda s}/[e^{-(i-1)h\lambda} - e^{-\lambda ih}]\,ds$$
$$= 1 + \lambda ih - e^{-\lambda ih}(1 + \lambda ih - \lambda h)/\lambda(1 - e^{\lambda h}) \tag{8.63}$$

Using Eq. (8.63), (8.57) can be simplified to:

$$E[L/A_i] = (1 + \lambda h - e^{\lambda h})/\lambda(1 - e^{-\lambda h}) \tag{8.64}$$

Finally, using Eq. (8.64), $E(L)$ can be written as:

$$E(L) = (1 - (1 + \lambda h)e^{-\lambda h})/\lambda(1 - e^{-\lambda h}) \tag{8.65}$$

Hence, the expected length of the out-of-control period is

$$E(B) - E(L) = [h/(1 - \beta)] - [(1 - (1 + \lambda h)e^{-\lambda h})/\lambda(1 - e^{-\lambda h})] \tag{8.66}$$

8.5.3   Time to Take a Sample Batch and Interpret the Results

In his paper, Duncan[3] assumes that this interval is a constant, g, times the sample size n. Let the length of this interval be U, which then is

$$U = g \times n \tag{8.67}$$

8.5.4 Time to Find the Assignable Causes and Fix Them

Duncan[3] assumes that this interval is a constant equal to D, as per assumption 7. Now the total expected length of one cycle is

$$E(CT) = 1/\lambda + h/(1-\beta) - [(1-(1+\lambda h)e^{-\lambda h})/(\lambda(1-e^{-\lambda h}))] + g \times n + D$$

(8.68)

8.5.5 Expected Costs per Cycle

Duncan includes the following costs in the total costs per cycle, $E(C)$:

1. Cost of sampling
2. Cost of searching for an assignable cause when none exists (false alarm)
3. Cost of running the process in its out-of-control state
4. Cost of checking and eliminating an assignable cause

Let us now derive expressions for these costs.

8.5.5.1 *Expected Cost of Sampling*

Duncan assumes that the cost of sampling consists of a fixed cost of $\$a_1$ (per sample batch), which is independent of the sample size, n, and a variable cost of $\$a_2$ (per item). The expected sampling cost per cycle = sampling cost per batch × expected number of batches per cycle, or

$$(a_1 + a_2 \times n) \times E[CT]/h$$

(8.69)

where $E(CT)$ is given in Eq. (8.68).

8.5.5.2 *Expected Cost of Searching for an Assignable Cause When None Exists or Expected Cost of False Alarms per Cycle*

Duncan[3] assumes that each time a false alarm occurs, a unit cost of $\$a_3'$ is incurred. Hence,

$E[\text{cost of false alarms/cycle}]$
$$= a'_3 \times \text{Expected number of false alarms/cycle}$$

(8.70)

Let M be the number of false alarms/cycle, then

$$\text{Expected number of false alarms/cycle} = E(M)$$

$$= \sum_{m=0}^{\infty} m \times P[M = m]$$

(8.71)

where $P[M = m]$ is the probability distribution of the number of false alarms per cycle. This probability distribution can be related to the distribution of the length of time interval during which the process is in control (S). The expected value of false alarms per cycle can be shown to be equal to

$$E(M) = \sum_{m=0}^{\infty} (m\alpha)P[mh < s < (m+1)h]$$

$$= \alpha \sum_{m=0}^{\infty} m \int_{mh}^{(m+1)h} \lambda e^{-\lambda s}\, ds \tag{8.72}$$

$$E(M) = \alpha \sum_{m=0}^{\infty} m[e^{-mh\lambda} - e^{-(m+1)h\lambda}]$$

$$= \alpha \sum_{m=0}^{\infty} m e^{-mh\lambda} - \alpha \sum_{m=0}^{\infty} m e^{-(m+1)h\lambda} \tag{8.73}$$

In Eq. (8.73),

$$\alpha \sum_{m=0}^{\infty} m e^{-mh\lambda} = \alpha \sum_{m=0}^{\infty} m(e^{-\lambda h})^m$$

$$= (\alpha e^{-\lambda h})/(1 - e^{-\lambda h})^2 \tag{8.74}$$

and

$$\alpha \sum_{m=0}^{\infty} m e^{-(m+1)h\lambda} = \alpha e^{-\lambda h} \sum_{m=0}^{\infty} m e^{-mh\lambda}$$

$$= \alpha e^{-\lambda h}[e^{-\lambda h}/(1 - e^{-\lambda h})^2 \tag{8.75}$$

Now, Eq. (8.73) is simplified to

$$E(M) = \alpha e^{-\lambda h}/(1 - e^{-\lambda h})^2 - \alpha e^{-\lambda h}[e^{-\lambda h}/(1 - e^{-\lambda h})^2]$$

$$= \alpha e^{-\lambda h}/(1 - e^{-\lambda h}) \tag{8.76}$$

Now, Eq. (8.70) can be written as:

$$E[\text{Cost of false alarms/cycle}] = a_3' \times \alpha e^{-\lambda h}/(1 - e^{-\lambda h}) \tag{8.77}$$

8.5.5.3 Expected Cost of Running Process in Out-of-Control State

Duncan[3] assumes that the unit cost of running the process in its out-of-control state is a_4/unit time; therefore,

$E[\text{cost of running the process in its out-of-control state}]$

$$= a_4 \times \text{Expected length of time during which the process}$$
$$\text{is run in its out-of-control state} \tag{8.78}$$

The expected length of time the process is run in its out-of-control state is segment 2 of the cycle, and as per Eq. (8.66) is

$$(h/(1 - \beta)) - [(1 - (1 + \lambda h)e^{-\lambda h})/\lambda(1 - e^{-\lambda h})]$$

Hence, Eq. (8.78) is

$E[\text{Cost of running the process in out-of-control state}]$

$$= a_4 \times \{(h/(1 - \beta)) - [(1 - (1 + \lambda h)e^{-\lambda h})/\lambda(1 - e^{-\lambda h})]\} \quad (8.79)$$

8.5.5.4 Expected Cost of Finding an Assignable Cause and Fixing It

Duncan[3] assumes that this cost is a constant and is equal to $\$a_3$. Now, the total expected cost per cycle, $E(C)$, can be written as the sum of the costs derived earlier, and is equal to

$$E(C) = \{a_1 + a_2 \times n\}E(CT)/h + a_3' \times \alpha e^{-\lambda h}/(1 - e^{-\lambda h})$$
$$+ a_4 \times \{h/(1 - \beta) - [(1 - (1 + \lambda h)e^{-\lambda h})/\lambda(1 - e^{-\lambda h})]\} + a_3 \quad (8.80)$$

where $E(CT)$ is given in Eq. (8.68).

The objective function to be minimized is the expected total cost per unit time, $E(TC)$, which is obtained by dividing the total expected cost per cycle, $E(C)$, as per Eq. (8.80), by the total expected length of one cycle, $E(CT)$, as per Eq. (8.68):

$$E(TC) = [\{a_1 + a_2 n\}E(CT)/h + a_3' \times \alpha e^{-\lambda h}/(1 - e^{-\lambda h}) + a_4 \times \{(h/(1 - \beta))$$
$$- [(1 - (1 + \lambda h)e^{-\lambda h})/\lambda(1 - e^{-\lambda h})]\} + a_3]/[1/\lambda + h/(1 - \beta)$$
$$- [(1 - (1 + \lambda h)e^{-\lambda h})/[\lambda(1 - e^{-\lambda h})] + g \times n + D] \quad (8.81)$$

Because of the complexity of this function and the fact that n is an integer, this problem cannot be solved by taking partial derivatives of the objective function with respect to each of the decision variables and setting them equal to 0. Numerical methods or trial-and-error methods are recommended.

There are many papers that deal with the economic design of control charts. Montgomery,[6] Svoboda,[11] and Ho and Case[4] have provided excellent reviews of the work done in this area. The economic design models are not popular among the practitioners because of the complexity of the models and the difficulties in estimating the cost parameters.[7]

8.6   Some Recent Developments

The traditional c and U charts assume that the underlying distribution of the number of defects per unit or batch is a Poisson distribution, but this assumption is not valid in real-life applications. Also, the control limits are derived assuming a normal approximation. To overcome these shortcomings of the c and U charts, Xie and Goh[12] proposed new control limits based on geometric distribution. Chan et al.[2] recently developed a new type of control chart, the Cumulative Quantity Control chart, because of the unsatisfactory performance of the c and U charts for monitoring high-yield processes with low defect rates. Recently, Philippe[8] developed a special EWMA type of chart for monitoring the range R.

8.7   References

1. *Statistical Process Control*, QS-9000, Reference Manual, Automotive Industry Action Group, Southfield, MI, 1995.
2. Chan, L.Y., Xie, M., and Goh, T.N., Cumulative quantity control charts for monitoring production processes, *Int. J. Prod. Processes*, 38, 397–408, 2000.
3. Duncan, A.J., The economic design of $\bar{X}$ charts used to maintain current control of a process, *J. Am. Stat. Assoc.*, 51, 228–242, 1956.
4. Ho, C. and Case, K.E., Economic design of control charts: a literature review for 1981–1991, *J. Qual. Technol.*, 26, 39–53, 1994.
5. Lucas, J.M. and Saccucci, M.S., Exponentially weighted moving average control schemes: properties and enhancements, *Technometrics*, 32, 1–12, 1990.
6. Montgomery, D.C., The economic design of control charts: a review and literature survey, *J. Qual. Technol.*, 14, 75–87, 1980.
7. Montgomery, D.C., *Introduction to Statistical Quality Control*, 3rd ed., John Wiley & Sons, New York, 1997.
8. Philippe, C., An R-EWMA Control Chart for Monitoring the Logarithm of the Process Range, working paper, Ecole des Mines de Nantes, France, 2000.
9. Roberts, S.W., Control chart tests based on geometric moving averages, *Technometrics*, 1, 239–250, 1959.
10. Ross, S.M., *Introduction to Probability Models*, 6th ed., Academic Press, San Diego, CA, 1997.
11. Svoboda, L., Economic design of control charts: a review and literature survey (1979–1989), in *Statistical Process Control in Manufacturing*, J.B. Keats and D.C. Montgomery, Eds., Marcel Dekker, New York, 1991.
12. Xie, M. and Goh, T.N., The use of probability limits for process control based on geometric distribution, *Int. J. Qual. Reliability Manage.*, 14, 64–73, 1997.

8.8 Problems

1. The in-control mean of the outside diameter of a component is 0.65" and the in-control standard deviation is 0.001". The diameters are normally distributed.

   a. What are the control limits of the $\bar{X}$ chart, if $\alpha = 0.0026$ and the sample size is 6?

   b. What is the probability of running the process when the mean shifts to 0.651"?

2. The in-control mean and the standard deviation of a quality characteristic are 2.50" and 0.005", respectively. The characteristic is normally distributed.

   a. What are the control limits of the $\bar{X}$ chart, if $\alpha = 0.005$ and $n = 6$?

   b. What fraction of components is expected to fall outside the tolerance limits of 2.500 ± 0.01" when the process is in control?

   c. What is the probability of Type II error if the mean shifts to 2.505"?

3. Sample batches of size $n = 5$ are collected from a process each day. After 25 sample batches have been collected, $\bar{\bar{X}} = 30$ and $\bar{s} = 2.1$. Both charts exhibit a process that is in control. The quality characteristic is normally distributed. Assume that $\alpha = 0.0026$.

   a. Estimate the process standard deviation.

   b. Determine the limits for the $\bar{X}$ and s charts.

4. A manufacturer is interested in controlling a process, using an $\bar{X}$ chart. Historical data for 5 sample batches, each consisting of two observations, are as follows.

Sample Batch	$\bar{X}$	Standard Deviation (s)
1	99	0.8
2	98	1.0
3	104	1.0
4	101	1.4
5	98	0.8

   Determine the control limits of the $\bar{X}$ and s charts, assuming $\alpha = 0.0026$.

5. The surface roughness values of ten successive surface plates ground on a grinding machine are as follows. Set up the limits for the individual and moving range charts. Assume that $\alpha = 0.0026$.

Item	Surface Roughness
1	41
2	42
3	44
4	42
5	40
6	42
7	41
8	43
9	41
10	40

6. The following are the hardness values of ten powdered metal components made out of successive blends. Set up the control limits for the appropriate control charts, with $\alpha = 0.0026$.

Blend	Hardness
1	65
2	63
3	61
4	64
5	66
6	64
7	62
8	60
9	61
10	64

7. The following data were collected from a foundry regarding the hardness of cast iron from successive charges of the cupola.

$$52 \quad 50 \quad 48 \quad 51 \quad 50$$

Assume that $\mu_0 = 50$, $\sigma_0 = 2$, $L = 3$, and $\lambda = 0.8$. Set up an EWMA control chart and compute the test statistics and control limits. Also, identify the out-of-control points, if any.

8. In order to monitor the quality of the picture tubes of television sets, the following data were collected. These are the actual number of nonconforming units found in sample batches of size 50 each:

Batch #	1	2	3	4	5	6	7	8	9	10
Nonconforming units	2	1	6	3	3	1	8	1	4	5

Set up the appropriate control chart(s) (with LCL, UCL, and CL) using the above data. Assume $\alpha = 0.0026$.

9. An automobile manufacturer wants to monitor the quality of the automobiles manufactured. The following data were collected on the number of nonconformities in 10 sample batches with 12 automobiles in each:

Batch #	1	2	3	4	5	6	7	8	9	10
Nonconformities	2	10	2	1	5	6	4	7	5	3

Set up the appropriate control chart(s) (with LCL, UCL, and CL) using the above data. Assume $\alpha = 0.0026$.

10. Aluminum sheets are produced in sizes of $4' \times 4'$. The producer is concerned about the surface defects on the sheets and collected the following data to monitor the production process:

Batch #	1	2	3	4	5	6	7	8	9	10
Batch size	1	5	2	3	2	2	3	1	1	3
Surface defects in the batch	1	2	0	4	2	0	6	0	1	1

Set up the appropriate control chart(s) (with LCL, UCL, and CL) using the above data. Assume $\alpha = 0.0026$.

11. Suppose that your assistant has been maintaining three c control charts: one to monitor the number of defects of type A per product (c_A), one to monitor the number of defects of type B per product (c_B), and the third chart to monitor the number of defects of type C per product (c_C). Let us assume that a type A defect costs your company $\$a_1$ to fix, a type B defect costs $\$a_2$ to fix, and a type C defect costs $\$a_3$. Suppose that you want to maintain one control chart to monitor the total amount in dollars required to fix all the defects of types A, B, and C per product. That is, the test statistic to be plotted on the new control chart is total cost per product = $a_1 c_A + a_2 c_B + a_3 c_C$, where c_A, c_B and c_C are the actual number of defects of types A, B, and C per product, respectively. Assume that the mean number of defects of types A, B, and C per product are C_{0A}, C_{0B}, and C_{0C}, respectively.

a. What are the limits of the new control chart? Assume that $\alpha = 0.005$. Assume also that c_A, c_B, and c_C are independent.

b. If $a_1 = \$5.00$, $a_2 = \$15.00$, and $a_3 = \$10.00$; the mean number of defects of type A per product (C_{0A}) = 3; the mean number of defects of type B per product (C_{0B}) = 4; and the mean number of defects of type C per product (C_{0C}) = 8, what are the control limits of this new chart? ($\alpha = 0.005$.)

12. An $\overline{X}$ chart is used to monitor the mean of a quality characteristic. The process is judged to be out of control if the process shifts by 1.5 standard deviations. The length of time the process is in control is exponentially distributed with a mean of 100 hours. The fixed sampling cost is $1.00 per batch and the variable cost is $0.20 per sample observation. It costs $10.00 to investigate a false alarm, $2.50 to find an assignable cause, and $50.00 per hour if the process is run in its out-of-control state. The time required to collect and evaluate a sample is 0.02 hour and it takes 1.5 hours to locate an assignable cause.

Assume that $h = 30$ minutes, $k = 2.81$, and $n = 5$. What is the expected cost per hour if this control chart is used?

9

Design of Experiments

CONTENTS

9.1 Introduction

In the design of experiments, a single experiment or a sequence of experiments is performed to test and quantify the effects of one or more input variables on one or more output variables of a process or a product. The design of experiments may be used to help improve the capability of a process by identifying the process and product variables that affect the mean and the variance of the quality characteristic(s) of the product. It can also help in improving process yields. The variables that affect the output variables are divided into two groups:

1. Input variables or signal factors
2. Noise variables

9.1.1 Input Variables or Signal Factors

Input variables or signal factors can be set at the desired levels by the experimenter; that is, these variables can be controlled during the experiment and at the design stage and/or in the actual production stage.

9.1.2 Noise Variables

Noise variable factors either cannot be controlled or are difficult and/or expensive to control during the design or actual production stage. Some examples of these factors are the composition of raw materials used in manufacture and the humidity level in a production shop. Both these variables could be controlled but only at considerable cost.

9.1.3   Other Variables

A third type of variable which includes variables that are functions of input variables and affect the output variables, is called an *intermediate variable*. The output variable that is being studied and measured in an experiment is the *response variable*. Identifying the input variables, intermediate variables, noise variables, and response variables is a critical step in any experiment and can be effectively performed using cause-and-effect diagrams.[1]

9.1.4   Replication

Experimental runs under identical conditions should be replicated a sufficient number of times to obtain accurate estimates of experimental errors and the effects of the input variables on the response variables(s).

9.1.5   Randomization

The order of assigning objects or components to the levels of factors and experimental runs should be randomized as much as possible in order to balance out the effect of the noise variables on the response variables, to minimize the bias, and to introduce independence among the observations of the response variables. This may not be easy or feasible in all experiments.

9.2   Single-Factor Experiments

In these experiments, the effect of only one signal factor or input variable on a response variable is studied.

9.2.1   Analysis

The following assumptions are made in this design.

1. The single factor is set at a different levels.
2. At each of the levels of the factor, n experiments are conducted; that is, the number of replications is n.
3. The value of the jth response variable at the ith level of the factor is denoted by y_{ij}, $i = 1, 2, ..., a$ and $j = 1, 2, ..., n$.
4. The y_{ij} for any given level i are independent and follow normal distribution with a mean μ_i and variance σ^2, $i = 1, 2, ..., a$.

Let us consider the following example.

TABLE 9.1

Data for Example 9.1

Temperature	Replication 1	Replication 2	Replication 3	Row Mean
600°F	5 (y_{11})	6 (y_{12})	7 (y_{13})	6.0 ($\bar{y}_{1.}$)
650°F	3 (y_{21})	4 (y_{22})	5 (y_{23})	4.0 ($\bar{y}_{2.}$)
700°F	7 (y_{31})	8 (y_{32})	9 (y_{33})	8.0 ($\bar{y}_{3.}$)

Example 9.1

A process engineer wants to test the effect of annealing temperature on the tensile strength of a component. She selects three levels of annealing temperature: 600°F, 650°F, and 700°F. A total of nine identical components are selected for the experiment, and three components are tested at each of the three levels. The components are randomly assigned to be annealed at the three levels of annealing temperature. The coded values of the tensile strengths of the nine components are given in Table 9.1. In this example, $a = 3$ and $n = 3$. The jth observation at level i of the factor, y_{ij}, can be written as:

$$y_{ij} = \mu_i + e_{ij} \tag{9.1}$$

where μ_i is the population mean of the observations at level i of the factor and e_{ij} is the associated error. Because of the assumption that the y_{ij} are independent and normally distributed with a mean μ_i and variance σ^2, the error terms, e_{ij}, are also independent and normally distributed with a mean 0 and a variance σ^2. The observation y_{ij} can also be written as:

$$y_{ij} = \mu + (\mu_i - \mu) + (y_{ij} - \mu_i) \tag{9.2}$$

where $\mu = \frac{\sum_{i=1}^{k}\mu_i}{k}$ is the grand mean.

Let

$$\tau_i = \mu_i - \mu \tag{9.3}$$

which represents the effect of level i of the factor on the response variable. From Eqs. (9.1) and (9.2),

$$e_{ij} = y_{ij} - \mu_i \tag{9.4}$$

Hence, Eq. (9.2) becomes:

$$y_{ij} = \mu + \tau_i + e_{ij}, \quad \text{for } i = 1, 2, \ldots, a \quad \text{and} \quad j = 1, 2, \ldots, n \tag{9.5}$$

This is the linear model for single-factor experiments. The sum of the squares of errors, e_{ij}, is

$$\sum_{i=1}^{a}\sum_{j=1}^{n}e_{ij}^2 = \sum_{i=1}^{a}\sum_{j=1}^{n}(y_{ij}-\mu_i)^2 \qquad (9.6)$$

It can be shown that the estimate of μ_i, which minimizes this quantity, is

$$\bar{y}_{i\cdot} = \frac{\sum_{j=1}^{n}y_{ij}}{n}, \quad i = 1, 2, \ldots, a \qquad (9.7)$$

As the e_{ij} are independent with mean $= 0$ and a common variance, the estimate given in Eq. (9.7) is also an unbiased estimator of μ_i, $i = 1, 2, \ldots, a$. Upon replacing e_{ij} and μ_i by the respective estimators in Eq. (9.6), we get:

$$\sum_{i=1}^{a}\sum_{j=1}^{n}e_{ij}^2 = \sum_{i=1}^{a}\sum_{j=1}^{n}(y_{ij}-\bar{y}_{i\cdot})^2 \qquad (9.8)$$

This is called the *sum of squares due to error* (SSE). The total variation present in the data is

$$\text{Total variation} = \sum_{i=1}^{a}\sum_{j=1}^{n}(y_{ij}-\bar{y}_{\cdot\cdot})^2 \qquad (9.9)$$

where

$$\bar{y}_{\cdot\cdot} = \frac{\sum_{i=1}^{a}\sum_{j=1}^{n}y_{ij}}{an} \qquad (9.10)$$

which is the estimate of μ. The total variation is called the *sum of squares total* (SST). The difference between SST and SSE is obtained as follows:

$$\text{SST} - \text{SSE} = \sum_{i=1}^{a}\sum_{j=1}^{n}(y_{ij}-\bar{y}_{\cdot\cdot})^2 - \sum_{i=1}^{a}\sum_{j=1}^{n}(y_{ij}-\bar{y}_{i\cdot})^2$$

$$= n\sum_{i=1}^{a}(\bar{y}_{i\cdot}-\bar{y}_{\cdot\cdot})^2. \qquad (9.11)$$

It can be seen that the term on the right-hand side is a positive quantity, hence SST $\geq$ SSE. Rearranging the terms, Eq. (9.11) can be written as:

$$\sum_{i=1}^{a}\sum_{j=1}^{n}(y_{ij}-\bar{y}_{\cdot\cdot})^2 = n\sum_{i=1}^{a}(\bar{y}_{i\cdot}-\bar{y}_{\cdot\cdot})^2 + \sum_{i=1}^{a}\sum_{j=1}^{n}(y_{ij}-\bar{y}_{i\cdot})^2 \qquad (9.12)$$

The term on the left-hand side of Eq. (9.12) is the sum of squares total (SST), the first term on the right-hand side is called the *treatment sum of squares* (SS treatment) and the second term is the sum of squares due to error (SSE). So, in the single-factor model, the total sum of squares is partitioned into the treatment sum of squares and the sum of squares due to error.

There is an alternate way of obtaining the relation in Eq. (9.12). The observation y_{ij} can be written as follows (replacing the means in Eq. (9.2) by their respective estimators):

$$y_{ij} = \bar{y}_{..} + (\bar{y}_{i.} - \bar{y}_{..}) + (y_{ij} - \bar{y}_{i.}) \tag{9.13}$$

which gives

$$y_{ij} - \bar{y}_{..} = (\bar{y}_{i.} - \bar{y}_{..}) + (y_{ij} - \bar{y}_{i.})$$
$$(y_{ij} - \bar{y}_{..})^2 = [(\bar{y}_{i.} - \bar{y}_{..}) + (y_{ij} - \bar{y}_{i.})]^2$$

and

$$\sum_{i=1}^{a}\sum_{j=1}^{n}(y_{ij} - \bar{y}_{..})^2 = \sum_{i=1}^{a}\sum_{j=1}^{n}[(\bar{y}_{i.} - \bar{y}_{..}) + (y_{ij} - \bar{y}_{i.})]^2$$
$$= \sum_{i=1}^{a}\sum_{j=1}^{n}(\bar{y}_{i.} - \bar{y}_{..})^2 + \sum_{i=1}^{a}\sum_{j=1}^{n}(y_{ij} - \bar{y}_{i.})^2$$
$$+ 2\sum_{i=1}^{a}\sum_{j=1}^{n}(\bar{y}_{i.} - \bar{y}_{..})(y_{ij} - \bar{y}_{i.}) \tag{9.14}$$

As the third term can be shown to be equal to zero, Eq. (9.14) becomes:

$$\sum_{i=1}^{a}\sum_{j=1}^{n}(y_{ij} - \bar{y}_{..})^2 = n\sum_{i=1}^{a}(\bar{y}_{i.} - \bar{y}_{..})^2 + \sum_{i=1}^{a}\sum_{j=1}^{n}(y_{ij} - \bar{y}_{i.})^2$$

which is the same as Eq. (9.12)

The original objective of testing the effect of the single factor on the response variable is equivalent to testing the following hypotheses:

Null hypothesis, H_0: $\mu_1 = \mu_2 = \cdots = \mu_k$.

Alternate hypotheses, H_1: At least two μ_i are unequal.

Now we need to develop an appropriate test statistic with which we can test these hypotheses.

The important underlying factor in developing the test statistic is the assumption that all the y_{ij} have the same variance (σ^2). If the alternate hypothesis is true, then there are a sample batches, each one from a different population with a different mean but the same variance. The sample variance from each of these sample batches is an unbiased estimator of the common variance, σ^2. The sample variance from the ith batch is

$$s_i^2 = \frac{\sum_{j=1}^{n}(y_{ij} - \bar{y}_{i\cdot})^2}{(n-1)}, \quad i = 1, 2, \ldots, a \tag{9.15}$$

The pooled estimator of σ^2, using these k estimators, is

$$\hat{\sigma}^2 = \frac{\sum_{i=1}^{a} s_i^2}{a} = \frac{\sum_{i=1}^{a}\sum_{j=1}^{n}(y_{ij} - \bar{y}_{i\cdot})^2}{a(n-1)}$$

$$= \frac{\text{SSE}}{a(n-1)} \tag{9.16}$$

The next step in the development of the test statistic is to divide the sums of squares by the respective degrees of freedom, which is the number of statistically independent elements in the associated sum of squares. Estimation of a parameter causes loss of one degree of freedom. The calculation of SST requires calculation of $\bar{y}_{\cdot\cdot}$, which estimates μ. Hence, the degrees of freedom associated with SST are the total number of observations -1:

$$\text{Degrees of freedom (SST)} = an - 1 = N - 1 \tag{9.17}$$

where N is the total number of observations. The degrees of freedom associated with SS treatments is the total number of levels of the factor used in the experiments -1:

$$\text{Degrees of freedom (SS treatments)} = a - 1 \tag{9.18}$$

The degrees of freedom (DF) associated with SSE is computed by subtracting the degrees of freedom for SS Treatments from the degrees for SST. That is,

$$\begin{aligned} DF(\text{SSE}) &= DF(\text{SST}) - DF(\text{SS treatments}) \\ &= (an - 1) - (a - 1) \\ &= a(n - 1) \\ &= N - a \end{aligned} \tag{9.19}$$

Dividing the sums of squares by the associated degrees of freedom yields the associated mean squares; that is,

$$\text{Mean square (treatments)} = \text{MST} = \text{SS(treatments)}/(a - 1) \tag{9.20}$$

$$\text{Mean square (error)} = \text{MSE} = \text{SSE}/a(n - 1) \tag{9.21}$$

The right-hand side of Eq. (9.21) is the same as the right-hand side of Eq. (9.16), hence

$$E(\text{MSE}) = \sigma^2 \qquad (9.22)$$

The expected value of MST can be shown to be equal to

$$E(\text{MST}) = \sigma^2 + n\theta^2 \qquad (9.23)$$

where

$$\theta^2 = \frac{\sum_{i=1}^{a}(\mu_i - \mu)^2}{(a-1)} \qquad (9.24)$$

which is a function of the difference among the means of the levels. This quantity will be zero if there is no difference among the means. Hence, both MSE and MST are unbiased estimators of the common variance σ^2 if there is no difference among the means, μ_i. If the means are different, then MSE is still an unbiased estimator of σ^2, but MST is a biased estimator of the common variance σ^2 because it estimates an additional quantity $n\theta^2$.

9.2.1.1 Some Results from Theory of Statistics

1. SSE/σ^2 follows a chi-square distribution with $a(n-1)$ degrees of freedom.
2. $\text{SS(treatments)}/\sigma^2$ follows a chi-square distribution with $(a-1)$ degrees of freedom if the means are equal; that is, if the null hypothesis, $H_0: \mu_1 = \mu_2 = \cdots = \mu_a$, is true.
3. The ratio $\dfrac{\text{SS(treatments)}/(a-1)}{\text{SSE}/[a(n-1)]} = \dfrac{\text{MST}}{\text{MSE}}$ follows an F distribution with $(a-1)$ numerator degrees of freedom and $a(n-1)$ denominator degrees of freedom when the null hypothesis is true.

From result 3, the logical test statistic is

$$\text{Computed } F = \frac{\text{MST}}{\text{MSE}} \qquad (9.25)$$

which is large if the alternate hypothesis—H_1: At least two μ_i are unequal—is true. This gives the following acceptance/rejection rules for the hypotheses testing for a specified probability of type I error, α:

Accept H_0 if the test statistic is less than or equal to $F_\alpha[(a-1), a(n-1)]$.
Otherwise, reject H_0 and accept H_1.

TABLE 9.2

ANOVA Table for Single-Factor Experiments

Source of Variation	Sum of Squares	Degrees of Freedom	Mean Square	Computed F (Test Statistic)	F Value from Table A.3
Treatments	SS (treatments)	$(a-1)$	$SS(T)/(a-1)$ $= MST$	MST/MSE	
Error	SSE	$a(n-1)$	$SSE/a(n-1)$ $= MSE$	—	
Total	SST	$an-1$	—	—	

The values of $F_\alpha[(a-1), a(n-1)]$ are tabulated in the F table given in Table A.3 in the Appendix for $\alpha = 0.05$ and 0.01. Calculation of the sums of squares, degrees of freedom, and mean squares is summarized in the analysis of variance (ANOVA) table given in Table 9.2.

The following computationally simpler formulas can be used to compute the sums of squares:

$$SST = \sum_{i=1}^{a}\sum_{j=1}^{n}(y_{ij}-\bar{y}_{..})^2 = \sum_{i=1}^{a}\sum_{j=1}^{n}y_{ij}^2 - \frac{y_{..}^2}{an} \qquad (9.26)$$

$$SS(\text{treatments}) = n\sum_{i=1}^{a}(\bar{y}_{i.}-\bar{y}_{..})^2 = \frac{\sum_{i=1}^{n}y_{i.}^2}{n} - \frac{y_{..}^2}{an} \qquad (9.27)$$

$SSE = SST - SS(\text{treatments})$. In Eqs. (9.26) and (9.27), $y_{i.} = \sum_{j=1}^{n}y_{ij}$, $i = 1, 2, \ldots a$ and $y_{..} = \sum_{i=1}^{a}\sum_{j=1}^{n}y_{ij}$.

9.2.1.2 Computation of Sum of Squares

$$SST = 5^2 + 6^2 + 7^2 + 3^2 + 4^2 + 5^2 + 7^2 + 8^2 + 9^2 - \frac{54^2}{3\times 3} = 30.0$$

$$SS(\text{treatment}) = \frac{18^2 + 12^2 + 24^2}{3} - \frac{54^2}{9} = 24.0$$

$$SSE = 30 - 24 = 6.0$$

Let us assume that $\alpha = 0.05$ (level of significance). The ANOVA results are given in Table 9.3. As $12.0 > 5.14$, we conclude that temperature affects tensile strength at $\alpha = 0.05$.

9.2.2 Confidence Intervals for the Treatment Means

Using the above results, the $100(1-\alpha)\%$ confidence interval for the mean for treatment i (level i of the single factor), μ_i, is

$$\left[\bar{y}_{i.} - t_\alpha[a(n-1)]\frac{s}{\sqrt{n}},\ \bar{y}_{i.} + t_\alpha[a(n-1)]\frac{s}{\sqrt{n}}\right] \qquad (9.28)$$

TABLE 9.3

ANOVA Table for Example 9.1

Source of Variation	Sum of Squares	Degrees of Freedom	Mean Square	Computed F (Test Statistic)	F Value from Table A.3 ($\alpha = 0.05$)
Temperature	24	$3 - 1 = 2$	$24/2 = 12$	$12/1 = 12.0^a$	5.14
Error	6	$8 - 2 = 6$	$6/6 = 1$	—	
Total	30	$9 - 1 = 8$	—	—	

[a] Indicates significance at $\alpha = 0.05$.

where $s = \sqrt{MSE}$ and $t_{\alpha/2}[a\,(n-1)]$ is the t value with $a(n-1)$ degrees of freedom and area to its right $= \alpha/2$. The t values can be found in Table A.2 in the Appendix.

Example 9.2

In Example 9.1, let $\alpha = 0.05$ (95% confidence interval). As $a = n = 3$, the degrees of freedom $= 3(3-1) = 6$. $t_{0.025}(6) = 2.4469$; $\sqrt{MSE} = \sqrt{1} = 1.0$. The limits for the mean of treatment 1 are $[6 - 2.4469 \times (1/\sqrt{3}),\ 6 + 2.4469 \times (1/\sqrt{3})] = (4.59, 7.41)$.

9.2.3 Fitting Equations

In most design of experiment problems, the experimenter is usually interested in obtaining a quantitative relationship between the response (dependent) variable and the independent variables. This quantitative relationship is usually expressed as a mathematical equation. For example, in an experiment with a single factor (independent variable), the relationship between the response variable y and the independent variable x could be expressed by the following linear equation, assuming a linear relationship:

$$y = \beta_0 + \beta_1 x, \tag{9.29}$$

where β_0 is the intercept on the vertical axis and β_1 is the slope of the line. These parameters are estimated using the data on y and x using regression analysis based on the method of least squares.

The linear model is restrictive and may not adequately capture the true relationship between y and x. The polynomial model is a better model to represent the relationship between y and x, assuming that x is a quantitative variable:

$$y = \beta_0 + \beta_1 x + \beta_2 x^2 + \beta_3 x^3 + \beta_4 x^4 + \cdots + \beta_p x^p + \varepsilon \tag{9.30}$$

which is a pth-degree polynomial (or the order of the polynomial is p).

- The most common objective in obtaining the relationship between y and x is to determine the lowest possible order polynomial equation (to minimize p) that still adequately describes the relationship.

- The order of a polynomial equation depends upon the number of levels of x. The maximum value of p = the number of levels −1.
- The coefficients of the polynomial equation can be determined by using regression methods.[4]
- If the levels of the factor (independent variable) are equally spaced, then determining the polynomial equation is simplified by using orthogonal polynomials.[3]
- If the levels of the factor are not equally spaced, then a regular polynomial equation can be used and the parameters can be determined using regression analysis (method of least squares), which software such as MINITAB and SAS performs efficiently.

9.2.4 Diagnosis Agreement Between the Data and the Model

The analysis of variance technique developed to test whether the effect of the factor on the response variable is significant depends upon the important assumption that the observations are independent and normally distributed with a common variance, σ^2. This implies that the errors, e_{ij}, are independent and normally distributed with a mean 0. In order to verify this assumption, tests must be performed on the data and the residuals, which estimate the errors:

$$\hat{e}_{ij} = y_{ij} - \bar{y}_{i\cdot} \quad \text{for all } i \text{ and } j \qquad (9.31)$$

(Refer to Example 9.1.) The residuals are given in Table 9.4. These residuals must be tested for independence by plotting the residuals against the treatment means for common variance using tests such as the Levene (Med) test, F Max test, and Bartlett's test[3] and for normality using tests such as the Kolmogorov–Smirnov tests. These diagnostic tests can be performed using MINITAB or SAS software. If the original data are not found to have homogeneity of variances, then transformation should be performed on the data so that the transformed data have equal variance.

The method of power transformations to stabilize variances uses the empirical relationship between the standard deviation and mean to obtain

TABLE 9.4

Residuals for Example 9.1

Temperature (°F)	Replication 1[a]	Replication 2[a]	Replication 3[a]	Row/Treatment Mean
600	(5) −1	(6) 0	(7) +1	6.0
650	(3) −1	(4) 0	(5) +1	4.0
700	(7) −1	(8) 0	(9) +1	8.0

[a] The original observations are enclosed within parentheses.

the transformation that results in homogeneity of variances.[2] Let us assume that in a particular experiment involving a single factor, a levels of the factor were tested with n replications at each level. Let the standard deviation and mean for level i be s_i and $\bar{y}_{i\cdot}$, respectively:

$$s_i = \sqrt{\frac{\sum_{j=1}^{a}(y_{ij} - \bar{y}_{i\cdot})^2}{n-1}}$$

$$\bar{y}_{i\cdot} = \frac{\sum_{j=1}^{n}y_{ij}}{n}$$

(The sample standard deviation s_i estimates the population standard deviation σ_i, and the sample mean $\bar{y}_{i\cdot}$ estimates the population mean μ_i, for $i = 1, 2, \ldots, a$.) As per the assumption for the linear model for a single-factor experiment, $\sigma_1 = \sigma_2 = \cdots = \sigma_a$.

Let us assume that in this example, the σ_i are not equal and that the relationship between the σ_i and the μ_i is of the form: $\sigma \propto \mu^\beta$, where β is a constant. Denoting the original variable (observations) by y, $\sigma_y \propto \mu^\beta$, which can be written as:

$$\sigma_y = \delta\mu^\beta \tag{9.32}$$

where δ is the proportionality constant.

Let us assume a power transformation of the type:

$$x = y^p$$

which will result in the following relation between the standard deviation and the mean of x:

$$\sigma_x \propto \mu_x^{p+\beta-1} \tag{9.33}$$

The condition to be satisfied for x to have homogeneity of variance is $p + \beta - 1 = 0$ or $p = 1 - \beta$. This requires estimation of the constant β, which can be obtained by replacing σ_y, δ, and μ by their respective estimators in Eq. (9.32):

$$s_i = \hat{\delta}\bar{y}_{i\cdot}^{\hat{\beta}} \tag{9.34}$$

Taking natural logarithms of both sides of Eq. (9.34) yields:

$$\ln(s_i) = \ln(\hat{\delta}) + \hat{\beta}\ln(\bar{y}_{i\cdot}) \tag{9.35}$$

This is a straight line, and the slope of the line is the estimate of β, which is $\hat{\beta}$. For example, if the estimated linear relationship is $\ln(s_i) = 0.85 + 0.75\ln(\bar{y}_{i\cdot})$, then $\hat{\beta} = 0.75$, and the required power transformation is $x = y^{1-0.75} = y^{0.25}$.

After the data are transformed, ANOVA should be performed and inferences made on the transformed data, but final conclusions must be stated with respect to the original data.

9.3 Random Effects Model

In Example 9.1, in which the effect of annealing temperature on the strength of the component was tested, the complete treatment population of interest consisted of the three levels (or treatments) of the annealing temperature (600°F, 650°F, and 700°F) only. These are called *fixed effect models*, in which most of the time inferences are made only on those particular treatments used in the experiment.

There are other types of studies in which the researcher would want to identify the major sources of variability in a population of treatments and estimate their variances. In such cases, the experimenter would randomly select the levels or treatments from the population. These are called *random effects models*. For example, if we are interested in testing whether there is a difference in the percentages of defective parts manufactured by the operators on the shop floor, we would select a certain number of operators randomly from a large group of operators on the shop floor and perform the experiment. This is a random effects model because a limited number of operators, who are considered levels or treatments, are selected randomly from a large population of operators and the results from this experiment will be extended to the entire population of the operators.

The model for the fixed effects model was given earlier in Eq. (9.5) as:

$$y_{ij} = \mu + \tau_i + e_{ij}$$

where $\tau_i = \mu_i - \mu$ is the effect of the ith treatment or level.

The model is the same for the random effects model. The major difference lies in τ_i, which is a constant in the fixed effects model. But, in the random effects model, τ_i is a random variable, because the treatments (levels) are chosen randomly and the inferences will be extended to the population of treatments.

Let us replace τ_i by f_i in the model, which then becomes:

$$y_{ij} = \mu + f_i + e_{ij}, \quad \text{for } i = 1, 2, ..., a \quad \text{and} \quad j = 1, 2, ..., n \qquad (9.36)$$

where f_i is the random effect for treatment i:

$$\mu_i = \mu + f_i, \quad \text{for } i = 1, 2, ..., a \qquad (9.37)$$

and $e_{ij} = y_{ij} - \mu_i$, as in the fixed effects model.

Let us assume:

1. The observation y_{ij} is assumed to follow a normal distribution with a mean of μ_i and a common variance, σ_y^2 for all i and j.

2. The random variables e_{ij} and f_i are assumed to be independent of each other.

3. The error e_{ij} is normally distributed with a mean of 0 and a variance of σ_e^2 for all i and j.

4. The random effect f_i is normally distributed with a mean of 0 and a variance of σ_f^2 for all i.

   • If $\sigma_f^2 = 0$, the population of treatments does not affect the response variable.

   • If $\sigma_f^2 > 0$, the population of treatments affects the response variable.

   • In this model, the treatment means, μ_i, are of no interest, because the treatments (or levels) are only a sample from the population of treatments. The parameter of interest is σ_f^2.

   • The variances σ_f^2 and σ_e^2 are called the *components of variance*, and the model in Eq. (9.36) is called the variance components model. Because of the assumptions made,

$$\sigma_y^2 = \sigma_e^2 + \sigma_f^2 \tag{9.38}$$

Let us look at the similarities and differences between the fixed effects model and random effects model.

Fixed Effects Model	Random Effects Model
$y_{ij} = \mu + \tau_i + e_{ij}$	$y_{ij} = \mu + f_i + e_{ij}$
$\text{Var}(y_{ij}) = \text{Var}(e_{ij})\,(\sigma_y^2 = \sigma_e^2)$	$\text{Var}(y_{ij}) = \sigma_y^2 = \sigma_f^2 + \sigma_e^2$
$H_0\colon \mu_1 = \cdots = \mu_a$	$H_0\colon \sigma_f^2 = 0$ (treatment does not affect response variable)
H_1: At least two μ_i are equal	$H_1\colon \sigma_f^2 > 0$ (treatment affects response variable)
SST = SS(treatments) + SSE	SST = SS(among group) + SS(within group)
	$\quad$ = SSA + SSE
$DF(\text{SST}) = a\,n - 1$	$DF(\text{SST}) = a\,n - 1$
$DF(\text{SS treatments}) = a - 1$	$DF(\text{SS among groups}) = D(\text{SSA}) = a - 1$
$DF(\text{SSE}) = a(n - 1)$	$DF(\text{SS within groups}) = a(n - 1)$
Mean square (treatments)	Mean square (among groups)
$= \text{MST} = \dfrac{\text{SS(treatments)}}{(a - 1)}$	$= \text{MSA} = \dfrac{\text{SSA}}{a - 1}$
Mean square (error) = $\text{MSE} = \dfrac{\text{SSE}}{a(n-1)}$	Mean square (within groups) = $\text{MSW} = \dfrac{\text{SSW}}{a(n-1)}$

$$E(\text{MST}) = \sigma_y^2 + n\theta, \quad \text{where } \theta = \frac{\Sigma(\mu_i - \mu)^2}{a - 1} \qquad E(\text{MSA}) = \sigma_e^2 + n\sigma_f^2 \tag{9.39}$$

$E(\text{MSE}) = \sigma_y^2$	$E(\text{MSW}) = \sigma_e^2$
Test statistic $= \dfrac{\text{MST}}{\text{MSE}}$	Test statistic $= \dfrac{\text{MSA}}{\text{MSW}}$

$$\text{Estimate of } \sigma_f^2 = \frac{\text{MSA} - \text{MSE}}{n} \tag{9.40}$$

Example 9.3

The following data show the effect of four operators chosen randomly on the number of defectives produced on a particular machine:

		Operators		
Lot	1	2	3	4
1	5	4	4	5
2	8	2	0	3
3	6	1	1	2
4	6	6	1	1
5	7	2	2	4
Total	32	15	8	15

TABLE 9.5

ANOVA for Example 9.3

Source of Variation	Sum of Squares	Degrees of Freedom	Mean Square	Computed F	F from Table A.3 ($\alpha = 0.05$)
Operators (group)	62.6	$4 - 1 = 3$	20.87	8.26[a]	3.24
Error	40.4	$19 - 3 = 16$	2.53	—	
Total	103.0	$20 - 1 = 19$	—	—	

[a] Indicates significance at $\alpha = 0.05$.

1. Perform an analysis of variance at the 0.05 level of significance.
2. Compute the estimates of the operator variance component and the experimental error variance component.

$$SST = 5^2 + 4^2 + \cdots + 2^2 + 4^2 - 70^2/20 = 103.0.$$
$$SSA = (32^2 + 15^2 + 8^2 + 15^2)/5 - 70^2/20 = 62.6.$$

The ANOVA results are given in Table 9.5. There is wide variation in the output among the operators at the 0.05 significance level (we are not selecting the best operator).

The estimate of the operator variance component = $\frac{20.87 - 2.53}{5} = 3.67$.

9.4 Two-Factor Experiments

9.4.1 Analysis

Even though study of single-factor experiments is necessary to understand the basis of ANOVA and other analyses, many experiments in real-life applications involve the study of the effects of two or more factors on the response variable at the same time. A design in which there is at least one observation for each of all possible combinations of all levels of all factors included in the study is called a *factorial design* or a *full factorial design*. The simplest type of factorial design is a two-factor experiment, in which the effects of two factors on one or more response variables are tested simultaneously.

In a two-factor experiment, the factors are usually denoted by A and B. Let the number of levels of factor A be a and the number of levels of factor B be b. Then, there is a total of $a \times b$ combinations in a full factorial design. If there are n replications, then there is a total of $n \times a \times b$ observations. The levels of factor A are $A_1, A_2, \ldots,$ and A_a, and the levels of factor B are $B_1, B_2, \ldots,$ and B_b.

9.4.1.1 Notation

- y_{ijk} = kth observation at level i of factor A and at level j of factor B, that is; the kth observation in cell (i, j), $i = 1, 2, \ldots, a$; $j = 1, 2, \ldots, b$; $k = 1, 2, \ldots, n$.

- Sum of observations at level i of factor $A = \sum_{j=1}^{b}\sum_{k=1}^{n}y_{ijk} = y_{i..}$

- Sum of observations at level j of factor $B = \sum_{i=1}^{a}\sum_{k=1}^{n}y_{ijk} = y_{.j.}$

- Sum of all the observations $= \sum_{i=1}^{a}\sum_{j=1}^{b}\sum_{k=1}^{n}y_{ijk} = y_{...}$

- Sample mean of observations at level i of factor $A = \frac{y_{i..}}{bn} = \bar{y}_{i..}$

- Sample mean of observations at level j of factor $B = \frac{y_{.j.}}{an} = \bar{y}_{.j.}$

- Grand mean of all observations $= \frac{y_{...}}{abn} = \bar{y}_{...}$

- Sum of observations in cell $(i, j) = \sum_{k=1}^{r}y_{ijk} = y_{ij.}$

- Sample mean of observations in cell $(i, j) = \frac{y_{ij.}}{n} = \bar{y}_{ij.}$

9.4.1.2 Assumptions

The observations in the cell (i, j) are assumed to be normally distributed with a mean of μ_{ij} and a common variance of σ^2. All the observations are independent of each other.

The mean of the (i, j)th cell = mean of the population at the ith level of A and jth level of $B = \mu_{ij}$:

$$\bar{\mu}_{i.} = \frac{\sum_{j=1}^{b}\mu_{ij}}{b} \qquad \bar{\mu}_{.j} = \frac{\sum_{i=1}^{a}\mu_{ij}}{a} \qquad \mu = \frac{\sum_{j=1}^{a}\sum_{j=1}^{b}\mu_{ij}}{ab}$$

The effects these means and their estimates are given in Table 9.6.

TABLE 9.6

Means and Their Estimates

Mean	Measures	Estimated by
μ_{ij}	Effects of level i of A and level j of B	$\bar{y}_{ij.}$
$\bar{\mu}_{i.}$	Effect of level i of A	$\bar{y}_{i..}$
$\bar{\mu}_{.j}$	Effect of level j of B	$\bar{y}_{.j.}$
μ	—	$\bar{y}_{...}$

TABLE 9.7

Data for Example 9.4

Temperature (A)	Composition (B)				Row Total ($y_{i.}$)	Row Mean $\bar{y}_{i..}$
	$j = 1$ (5%)	$j = 2$ (8%)	$j = 3$ (11%)	$j = 4$ (14%)		
$i = 1$ 600°F	5/7	3/5	8/4	7/5	44 ($y_{1.}$)	5.5
	12 (6)	8 (4)	12 (6)	12 (6)		
$i = 2$ 650°F	3/3	4/8	2/3	4/5	32 ($y_{2.}$)	4.0
	6 (3)	12 (6)	5 (2.5)	9 (4.5)		
$i = 3$ 700°F	7/6	5/6	7/9	6/6	52 ($y_{3.}$)	6.5
	13 (6.5)	11 (5.5)	16 (8)	12 (6)		
Column total ($y_{.j}$)	31 ($y_{.1}$)	31 ($y_{.2}$)	33 ($y_{.3}$)	33 ($y_{.4}$)	128 ($y_{..}$)	—
Column mean $\bar{y}_{.j.}$	5.17 $\bar{y}_{.1.}$	5.17 $\bar{y}_{.2.}$	5.5 $\bar{y}_{.3.}$	5.5 $\bar{y}_{.4.}$	—	5.33 $\bar{y}_{...}$

Note: In the (i, j) cells, the numbers underlined are the cell subtotals, y_{ij}, and the numbers within parentheses are the cell means, $\bar{y}_{ij.}$.

Example 9.4

Let us modify the single factor experiment in Example 9.1 to include material composition as the second factor, B, in addition to the annealing temperature, A. There are four levels of B ($b = 4$) and three levels of factor A ($a = 3$). Each of the 12 combinations is replicated two times ($n = 2$). The observations are given in Table 9.7.

In a two-factor experiment, in addition to the main effects (due to A and B), there is one other effect present. Consider the observations and their sample means in cells (1, 1), (1, 2), (2, 1), and (2, 2), at levels 1 and 2 of A and B, which are given below:

	Factor B Level 1	Factor B Level 2
Factor A Level 1	6	4
Factor A Level 2	3	6

Figure 9.1 contains the plot of these averages. It can be seen that the effect of A (annealing temperature) on the response variable (strength) depends upon the level of B (material composition). More specifically:

- At B_1 (5%), $\bar{y}_{i..}$ decreases from 6 to 3, as A increases from A_1 to A_2.
- At B_2 (8%), $\bar{y}_{i..}$ increases from 4 to 6, as A increases from A_1 to A_2.

This difference in the effect of A on the mean (of the response variable) for different levels of B is due to the interaction between A and B. If there is no interaction in this example, then the plot of the true means will be as given in

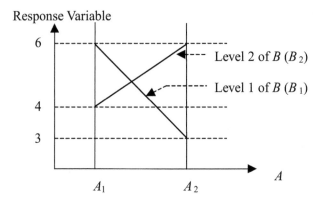

FIGURE 9.1
Presence of interaction effects.

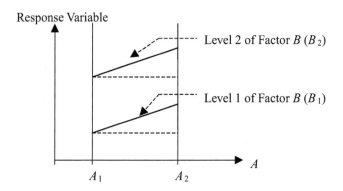

FIGURE 9.2
Absence of interaction effects.

Figure 9.2. Therefore, in a two-factor experiment, we can test the main effects of A and B and the interaction between A and B.

Let α_i represent the effect of level i of factor A and β_j the effect of level j of factor B on the response variable. Then,

$$\alpha_i = \bar{\mu}_{i\cdot} - \mu \qquad (9.41)$$

and its estimate is

$$\hat{\alpha}_i = \bar{y}_{i\cdot\cdot} - \bar{y}_{\cdot\cdot\cdot} \qquad (9.42)$$

and

$$\beta_j = \bar{\mu}_{\cdot j} - \mu \qquad (9.43)$$

and its estimate is

$$\hat{\beta}_i = \bar{y}_{\cdot j\cdot} - \bar{y}_{\cdots}$$ (9.44)

As μ_{ij} measures the combined effects of means $\bar{\mu}_{i\cdot}$ and $\bar{\mu}_{\cdot j}$ on the response variable, it can be written as:

$$\mu_{ij} = \mu + (\bar{\mu}_{i\cdot} - \mu) + (\bar{\mu}_{\cdot j} - \mu)$$ (9.45)

if there is no interaction between level i of factor A and level j of factor B. On the other hand, if there is interaction between level i of factor A and level j of factor B, it has to be the difference between the left-hand and right-hand sides of Eq. (9.45):

$$(\alpha\beta)_{ij} = (\mu_{ij} - \mu) - (\bar{\mu}_{i\cdot} - \mu) - (\bar{\mu}_{\cdot j} - \mu)$$
$$= \mu_{ij} - \bar{\mu}_{i\cdot} - \bar{\mu}_{\cdot j} + \mu$$ (9.46)

where $(\alpha\beta)_{ij}$ denotes the interaction between level i of A and level j of B. This is estimated by:

$$\widehat{(\alpha\beta)}_{ij} = \bar{y}_{ij\cdot} - \bar{y}_{i\cdot\cdot} - \bar{y}_{\cdot j\cdot} + \bar{y}_{\cdots}$$ (9.47)

It can be seen that a reasonable linear model in the two-factor experiment is

$$y_{ijk} = \mu + \alpha_i + \beta_j + (\alpha\beta)_{ij} + e_{ijk}$$ (9.48)

where e_{ijk} is the error equal to $y_{ijk} - \mu_{ij}$ and its estimate is $(y_{ijk} - \bar{y}_{ij\cdot})$.

9.4.1.3 Partitioning of the Total Sum of Squares

In two-factor experiments, the observation y_{ijk} can be written as:

$$y_{ijk} = y_{ijk} - \bar{y}_{ij\cdot} + \bar{y}_{ij\cdot} - \bar{y}_{i\cdot\cdot} + \bar{y}_{i\cdot\cdot} - \bar{y}_{\cdot j\cdot} + \bar{y}_{\cdot j\cdot} - \bar{y}_{\cdots} + \bar{y}_{\cdots} - \bar{y}_{\cdots} + \bar{y}_{\cdots}$$
$$(y_{ijk} - \bar{y}_{\cdots}) = (y_{ijk} - \bar{y}_{ij\cdot}) + (\bar{y}_{i\cdot\cdot} - \bar{y}_{\cdots}) + (\bar{y}_{\cdot j\cdot} - \bar{y}_{\cdots}) + (\bar{y}_{ij\cdot} - \bar{y}_{i\cdot\cdot} - \bar{y}_{\cdot j\cdot} + \bar{y}_{\cdots})$$ (9.49)
$$= \hat{e}_{ijk} + \hat{\alpha}_i + \hat{\beta}_j + \widehat{(\alpha\beta)}_{ij}$$

Hence,

$$\sum_{i=1}^{a}\sum_{j=1}^{b}\sum_{k=1}^{n}(y_{ijk} - \bar{y}_{\cdots})^2 = \sum_{i=1}^{a}\sum_{j=1}^{b}\sum_{k=1}^{n}(y_{ijk} - \bar{y}_{ij\cdot})^2 + bn\sum_{i=1}^{a}(\bar{y}_{i\cdot\cdot} - \bar{y}_{\cdots})^2$$
$$+ an\sum_{j=1}^{b}(\bar{y}_{\cdot j\cdot} - \bar{y}_{\cdots})^2 + n\sum_{i=1}^{a}\sum_{j=1}^{b}(\bar{y}_{ij\cdot} - \bar{y}_{i\cdot\cdot} - \bar{y}_{\cdot j\cdot} + \bar{y}_{\cdots})^2$$ (9.50)

FIGURE 9.3
Partitioning of SST.

All the other terms are equal to zero. In Eq. (9.50), the expression on the left-hand side is the sum of squares total (SST), the first term on the right-hand side is the sum of squares due to error (SSE), the second term is the sum of squares due to factor A (SSA), the third term is the sum of squares due to factor B (SSB), and the last term is the sum of squares due to the interaction between A and B (SSAB). Hence, Eq. (9.50) can be written as:

$$SST = SSE + SSA + SSB + SSAB \qquad (9.51)$$

This partitioning of SST is shown in Figure 9.3.

Simpler formulas for computing the sums of squares are given below:

$$SST = \sum_{i=1}^{a}\sum_{j=1}^{b}\sum_{k=1}^{n} y_{ijk}^2 - \frac{y_{...}^2}{abn} \qquad (9.52)$$

$$SSA = \frac{\sum_{i=1}^{a} y_{i..}^2}{bn} - \frac{y_{...}^2}{abn} \qquad (9.53)$$

$$SSB = \frac{\sum_{i=1}^{b} y_{.j.}^2}{an} - \frac{y_{...}^2}{abn} \qquad (9.54)$$

$$SSAB = \frac{\sum_{i=1}^{a}\sum_{j=1}^{b} y_{ij.}^2}{n} - \frac{y_{...}^2}{abn} - SSA - SSB \qquad (9.55)$$

$$SSE = SST - (SSA + SSB + SSAB) \qquad (9.56)$$

The hypotheses tested in a two-factor experiment are as follows:

- Effect of factor A:
  H_0: $\alpha_1 = \alpha_2 = \cdots = \alpha_a$.
  H_1: $\alpha_i \neq \alpha_j$ for some i and j.
- Effect of factor B:
  H_0: $\beta_1 = \beta_2 = \cdots = \beta_b$.
  H_1: $\beta_i \neq \beta_j$ for some i and j.

TABLE 9.8

Summary of Test Statistics

Sum of Squares	Degrees of Freedom	Mean Square	Test Statistic (Critical Value)
SST	$abn - 1$	—	—
SSA (effect of A)	$a - 1$	MSA $= \text{SSA}/(a-1)$	MSA/MSE $(F_\alpha[(a-1), ab(n-1)])$
SSB (effect of B)	$b - 1$	MSB $= \text{SSB}/(b-1)$	MSB/MSE $(F_\alpha[(b-1), ab(n-1)])$
SSAB (effect of interaction between A and B)	$(a-1)(b-1)$	MS(AB) $= \text{SSAB}/((a-1)(b-1))$	MSAB/MSE $(F_\alpha[(a-1)(b-1), ab(n-1)])$
SSE	$ab(n-1)$	MSE $= \text{SSE}/(ab(n-1))$	—

TABLE 9.9

ANOVA for Example 9.4

Source of Variation	Sum of Squares	Degrees of Freedom	Mean Square	Computed F	F from Table A.3 $(\alpha = 0.05)$
Temperature (A)	25.33	$3 - 1 = 2$	12.67	5.84[a]	3.89
Composition (B)	0.67	$4 - 1 = 3$	0.22	<1	
Interaction (AB)	27.33	6	4.56	2.10	3.49
Error	26.00	$23 - 11 = 12$	2.17	—	
Total	79.33	$24 - 1 = 23$	—	—	

[a] Indicates significance at $\alpha = 0.05$.

- Effect of Interaction between A and B:

  H_0: All $(\alpha\beta)_{ij} = 0$.

  H_1: At least one $(\alpha\beta)_{ij} \neq 0$.

The degrees of freedom, mean squares, and test statistics are summarized in Table 9.8. Let us now return to Example 9.4 and the necessary calculations:

$$\text{SST} = 5^2 + 7^2 + \cdots + 6^2 + 6^2 - \frac{128^2}{24} = 79.33.$$

$$\text{SSA} = \frac{44^2 + 32^2 + 52^2}{4 \times 2} - \frac{128^2}{24} = 25.33.$$

$$\text{SSB} = \frac{31^2 + 31^2 + 33^2 + 33^2}{3 \times 2} - \frac{128^2}{24} = 0.67.$$

$$\text{SSAB} = \frac{12^2 + 8^2 + 12^2 + \cdots + 16^2 + 12^2}{2} - \frac{128^2}{24} - 25.33 - 0.67 = 27.33.$$

$$\text{SSE} = 79.33 - (25.33 + 0.67 + 27.33) = 26.$$

The ANOVA results are given in Table 9.9. The effect of temperature on tensile strength is significant at $\alpha = 0.05$. The effects of composition (B) and interactions are not significant.

9.4.2 Without Replications

Example 9.5

Let us assume that in Example 9.4 the experimenter could not replicate the experiment and there is only one observation for each of the 12 combinations of the levels of A and B. Table 9.10 contains the observations. Necessary calculations include:

$$\text{SST} = 5^2 + 3^2 + \cdots + 7^2 + 6^2 - \frac{61^2}{12} = 40.92.$$

$$\text{SSA} = \frac{23^2 + 13^2 + 25^2}{4} - \frac{61^2}{12} = 20.67.$$

$$\text{SSB} = \frac{15^2 + 12^2 + 17^2 + 17^2}{3} - \frac{61^2}{12} = 5.58.$$

$$\text{SSAB} = 5^2 + 3^2 + \cdots + 7^2 + 6^2 - \frac{61^2}{12} - 20.67 - 5.58 = 14.67.$$

$$\text{SSE} = 40.92 - (20.67 + 5.58 + 14.67) = 0.$$

We need SSE and MSE to perform ANOVA. Here, there is no alternative except to assume that the interaction between A and B is negligible and to use SSAB as SSE. The resulting ANOVA is given in Table 9.11.

TABLE 9.10

Data for Example 9.5

Temperature (A)	Composition (B)				Row Total
	5%	8%	11%	14%	
600°F	5	3	8	7	23
650°F	3	4	2	4	13
700°F	7	5	7	6	25
Column total	15	12	17	17	61

TABLE 9.11

ANOVA for Example 9.5

Source of Variation	Sum of Squares	Degrees of Freedom	Mean Square	Computed F	F from Table A.3 ($\alpha = 0.05$)
Temperature (A)	20.67	2	10.34	4.22	5.14
Composition (B)	5.58	3	1.86	<1	—
Interaction (AB)	—	—	—	—	—
Error	14.67	6	2.45	—	—
Total	40.92	11	—	—	—

None of the effects is significant at $\alpha = 0.05$. The advantage of having more than one replication can be seen in this example. At least two replications for at least one combination are necessary for obtaining an explicit sum of square due to error.

9.4.3 Approximate Percentage Contributions of Effects and Error

The following formulas are used for computing the approximate percentage contribution of each effect (main factor or interaction) to the variability of the response variable:[7]

$$\% \text{ contribution of } U = \frac{MSU - MSE}{SST} \times DF(U) \times 100 \qquad (9.57)$$

where $U = A$, B, or AB (any main factor or interaction). This percentage is set to 0 if MSU < MSE.

$$\% \text{ contribution of error} = 100 - (\text{sum of } \% \text{ contributions of all} \\ \text{main factors and interactions}) \qquad (9.58)$$

In Example 9.4,

$$A = \frac{12.67 - 2.17}{79.33} \times 2 \times 100 = 26.5\%.$$

$$B = \frac{0.22 - 2.17}{79.33} \times 2 \times 100 < 0 \rightarrow 0\%.$$

$$AB = \frac{4.56 - 2.17}{79.33} \times 6 \times 100 = 18.0\%.$$

Error $= 100 - [26.5 + 18] = 55.5\%.$

A large percentage of contribution by error indicates possible changes in the levels of signal factors not controlled during the experiment and/or large inherent variation in the process due to noise variables.

9.4.4 Confidence Intervals

$100(1 - \alpha)\%$ confidence interval for the mean of response variable at level i of factor A:

$$\bar{y}_{i\cdots} \pm t_{\frac{\alpha}{2}}[ab(n-1)]\sqrt{\frac{MSE}{bn}} \qquad (9.59)$$

100(1 − α)% confidence interval for the mean of response variable at level j of factor B:

$$\bar{y}_{\cdot j\cdot} \pm t_{\frac{\alpha}{2}}[ab(n-1)]\sqrt{\frac{\text{MSE}}{an}} \tag{9.60}$$

100(1 − α)% confidence interval for the cell mean of response variable at level i of A and level j of factor B:

$$\bar{y}_{ij\cdot} \pm t_{\frac{\alpha}{2}}[ab(n-1)]\sqrt{\frac{\text{MSE}}{n}} \tag{9.61}$$

9.5   Nested Factor Designs

9.5.1   Description

In standard factorial treatment designs, each level of every factor occurs with all levels of the other factors. In such designs with more than one replication, all the interaction effects can be studied. In a nested design, the levels of one factor (for example, factor B) are similar but *not* identical for different levels of another factor (say, A). These are also called *hierarchical designs*.

Example 9.7

A company is interested in testing whether there is any difference among the percentage of defects produced on the three machines (1, 2, and 3) on the shop floor. They use a nested design with six operators—b_1, b_2, b_3, b_4, b_5, and b_6—who operate the machines and two replicates. (See Figure 9.4.) Observation y_{ijk} is the kth replicate, $k = 1$ and 2, on machine i, $i = 1, 2$, and 3, with operator j, $j = b_1$, b_2, b_3, b_4, b_5, and b_6. This is a two-stage nested design. If there are an equal number of levels of B within each level of A and an equal number of replicates, then the design is a balanced nested design. The effects that can be tested in this design are the effect due to machines (factor A) and the effect of operators nested within the machines (B/A). The error term is nested within levels of A and B. In this design, the interaction between A and B cannot be tested because every level of B does not appear with every level of A.

The linear model for this design is

$$y_{ijk} = \mu + \alpha_i + \beta_{j(i)} + e_{k(ij)} \tag{9.62}$$

where μ is the grand mean, α_i is the effect of level i of factor A, $\beta_{j(i)}$ is the effect of level j of factor B nested within level i of factor A, and $e_{k(ij)}$ is the error nested

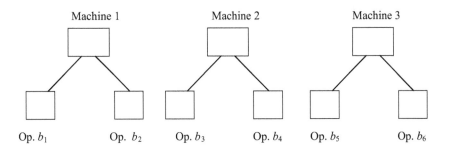

FIGURE 9.4
Nest design with two factors.

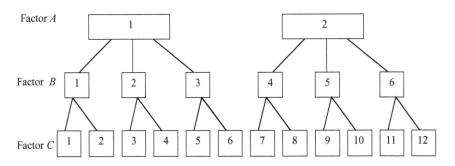

FIGURE 9.5
Nested factor design with three factors.

within levels i of A and j of B. Replacing the parameters in Eq. (9.62) by their respective estimators yields:

$$y_{ijk} = \bar{y}_{...} + (\bar{y}_{i..} - \bar{y}_{...}) + (\bar{y}_{ij.} - \bar{y}_{i..}) + (y_{ijk} - \bar{y}_{ij.}) \qquad (9.63)$$

We assume that the observations y_{ijk} are independent and normally distributed with a mean of μ_{ij} and a variance of σ^2.

An example of three-stage nested design is shown in Figure 9.5.

9.5.2 Analysis

9.5.2.1 Sums of Squares

Let the number of levels of factor A be a, the number of levels of B nested under each level of A be b, and the number of replications be n. The sums of squares and the associated degrees of freedom are as follows.

The total sum of squares is

$$SST = \sum_{i=1}^{a}\sum_{j=1}^{b}\sum_{k=1}^{n}(y_{ijk} - \bar{y}_{...})^2 = \sum_{i=1}^{a}\sum_{j=1}^{b}\sum_{k=1}^{n}y_{ijk}^2 - \frac{y_{...}^2}{abn} \qquad (9.64)$$

and it has $(abn - 1)$ degrees of freedom.

The sum of squares due to factor A is

$$\text{SSA} = bn \sum_{i=1}^{a} (\bar{y}_{i\cdot\cdot} - \bar{y}_{\cdots})^2 = \frac{\sum_{i=1}^{a} y_{i\cdot\cdot}^2}{bn} - \frac{y_{\cdots}^2}{abn} \tag{9.65}$$

and it has $(a - 1)$ degrees of freedom.
    The sum of squares due to B nested within A is

$$\text{SS(B/A)} = n \sum_{i=1}^{a} \sum_{j=1}^{b} (\bar{y}_{ij\cdot} - \bar{y}_{i\cdot\cdot})^2 = \frac{\sum_{i=1}^{a} \sum_{j=1}^{b} y_{ij\cdot}^2}{n} - \frac{y_{i\cdot\cdot}^2}{bn} \tag{9.66}$$

and it has $a(b - 1)$ degrees of freedom.
    The sum of squares due to error is computed as:

$$\text{SSE} = \text{SST} - [\text{SSA} + \text{SS(B/A)}] \tag{9.67}$$

which has $ab(n - 1)$ degrees of freedom.
    The mean squares and test statistics are obtained in a manner similar to those used in earlier designs.
    Let us add data to Example 9.6 (see Figure 9.6). Analyze the data and draw conclusions at a 5% significance level ($a = 3$, $b = 2$, $n = 2$).

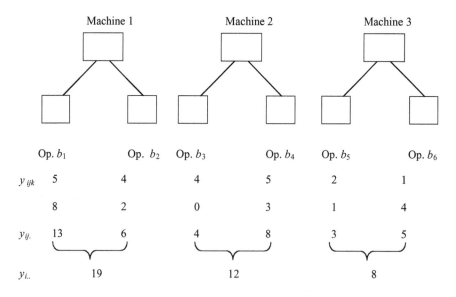

FIGURE 9.6
Data for Example 9.6.

TABLE 9.12

ANOVA for Example 9.6

Source of Variation	Sum of Squares	Degrees of Freedom	Mean Square	Computed F	F from Table A.3 ($\alpha = 0.05$)
Machines	15.5	2	7.75	2.16	5.14
Operators within machines	17.25	3	5.75	1.61	4.76
Error	21.5	6	3.58	—	
Total	54.25	12 − 1 = 11	—	—	

9.5.2.2 Calculations

$$SST = 5^2 + 8^2 + 4^2 + 2^2 + \cdots + 1^2 + 4^2 - \frac{39^2}{12} = 54.25.$$

$$SSA = \frac{19^2 + 12^2 + 8^2}{4} - \frac{39^2}{12} = 15.5.$$

$$SS(B/A) = \frac{13^2 + 6^2 + 4^2 + 8^2 + 3^2 + 5^2}{2} - \frac{19^2 + 12^2 + 8^2}{4} = 17.25.$$

The ANOVA results are given in Table 9.12. There is no difference among the number of defectives generated by the machines. Also, there is no difference among operators nested within each machine. In a nested experiment, the factors tested could be fixed or random or a combination of both. While, the computation of the sums of squares and the test statistics do not change whether these factors are fixed or random, the interpretation of the results depend upon the types of factors.

9.5.3 Staggered Nested Designs to Equalize Information About Variances

The nested factor design contains more information on factors at lower levels in the hierarchy of the design than at higher levels because of the larger degrees of freedom. In larger studies, the discrepancies in degrees of freedom among sources of variation can be considerable. Staggered nested designs were developed to equalize the degrees of freedom for the mean squares at each level of the hierarchy. The staggered designs have unequal numbers of levels for factors that are nested within other factors. The levels for factor B nested within factor A are varied from one level of factor A to another in such a way that the degrees of freedom for MSA and MS (B/A) are more equal.

Example 9.7

A staggered nested design is given in Figure 9.7. The degrees of freedom (machine) = 2. For three operators/machines, the degrees of freedom (operators/machines) = 2 + 2 + 2 = 6. In the above design, the degrees of freedom (operators/machines) = 2 + 1 + 1 = 4.

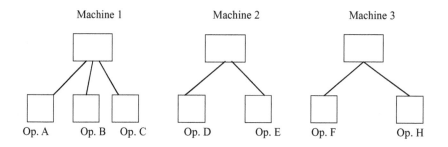

FIGURE 9.7
Staggered nested design.

9.6 2^k Factorial Experiments

9.6.1 Analysis

As the number of experiments (combinations) in a full factorial experiment is very large when the number of factors included in the experiment increases, it is common to use two levels for each factor studied. If the effect of a factor on the response variable is nonlinear, then selecting only two levels might miss the minimum or maximum effect. If there are k factors set at two levels each, then the total number of combinations is 2^k; therefore, these designs are called 2^k designs. The two levels of a factor are called *low* and *high*, even though this notation may be arbitrary in the case of qualitative factors (such as cleaning method or type of raw material). The other notation used to denote the levels are

Low	High
−	+
1	2

Example 9.8

An experiment with three factors is conducted to test the effect of wing length (A), body width (B), and tail length (C) on the descent time of a paper helicopter. The levels of these three factors are as follows:

	Low	High
Wing length (A)	1" (−)	$1^{1}/_{12}$" (+)
Body width (B)	1" (−)	$1^{3}/_{8}$" (+)
Tail length (C)	1" (−)	$1^{1}/_{2}$" (+)

TABLE 9.13

Data for Example 9.8

A (inches)	B (inches)	C (inches)	Replication 1	Replication 2	Total
−1	−1	−1	97	84	181
−1	−1	$+1^{1}/_{2}$	63	66	129
−1	$+1^{3}/_{8}$	−1	96	104	200
−1	$+1^{3}/_{8}$	$+1^{1}/_{2}$	72	72	144
$+1^{1}/_{2}$	−1	−1	87	100	187
$+1^{1}/_{2}$	−1	$+1^{1}/_{2}$	75	88	163
$+1^{1}/_{2}$	$+1^{3}/_{8}$	−1	104	100	204
$+1^{1}/_{2}$	$+1^{3}/_{8}$	$+1^{1}/_{2}$	59	81	160

TABLE 9.14

Rearranged Data for Example 9.8

A	B	C	AB	AC	BC	ABC	Obs. 1	Obs. 2	Total
−	−	−	+	+	+	−	97	84	181
−	−	+	+	−	−	+	63	66	129
−	+	−	−	+	−	+	96	104	200
−	+	+	−	−	+	−	72	72	144
+	−	−	−	−	+	+	87	100	187
+	−	+	−	+	−	−	75	88	163
+	+	−	+	−	−	−	104	100	204
+	+	+	+	+	+	+	59	81	140
									1348

The experiment was replicated twice, and the data are given in Table 9.13. The possible effects that can be studied in this experiment are the main factors A, B, and C and interactions AB, AC, BC, and ABC. The linear model for this design is

$$y_{ijkl} = \mu + \alpha_i + \beta_j + \gamma_k + (\alpha\beta)_{ij} + (\alpha\gamma)_{ik} + (\beta\gamma)_{jk} + (\alpha\beta\gamma)_{ijk} + e_{ijkl} \qquad (9.68)$$

In Eq. (9.68), y_{ijkl} is the *l*th observation at the *i*th level of A, *j*th level of B, and *k*th level of C. Also, μ is the grand mean; α_i, β_j, and γ_k are, respectively, the effects of level *i* of main factor A, level *j* of main factor B, and level *k* of main factor C; $(\alpha\beta)_{ij}$, $(\alpha\gamma)_{ik}$, and $(\beta\gamma)_{jk}$ are, respectively, the effects of the appropriate levels of the two factor interactions AB, AC, and BC; $(\alpha\beta\gamma)_{ijk}$ is the effect of the appropriate level of the three-factor interaction ABC; and e_{ijkl} is the error.

The observations are given in Table 9.14 with "−" representing low levels and "+" representing high levels. The signs in an interaction column are obtained by multiplying the signs in the associated columns of the main factors.

The sum of squares total (SST) is computed using the same formula used earlier. Because, in this design, each factor is set at two levels only, the formulas for the sum of squares of the factors are simpler:

$$SS(\text{any factor}) = \frac{(\text{Sum of the observations at ''--'' level})^2}{(\text{Number of observations at ''--'' level})}$$

$$+ \frac{(\text{Sum of the observations at ''+'' level})^2}{(\text{Number of observations at ''+'' level})}$$

$$- \frac{(\text{Grand sum of all the observations})^2}{\text{Total number of observations}}.$$

$$SST = 97^2 + 84^2 + \cdots + 59^2 + 81^2 - \frac{1348^2}{16} = 3417.$$

$$SSA = \frac{(187 + 163 + 204 + 140)^2}{8} + \frac{(129 + 181 + 144 + 200)^2}{8} - \frac{1348}{16} = 100.$$

$$SSB = \frac{(144 + 200 + 204 + 140)^2}{8} + \frac{(129 + 181 + 187 + 163)^2}{8} - \frac{1348^2}{16} = 49.00.$$

$$SSC = \frac{(129 + 144 + 163 + 140)^2}{8} + \frac{(181 + 200 + 187 + 204)^2}{8} - \frac{1348^2}{16} = 2401.00.$$

The sum of squares of the interactions are computed by considering "--" as one level and "+" as the second level:

$$SS(AB) = \frac{(129 + 181 + 204 + 140)^2}{8} + \frac{(144 + 200 + 187 + 163)^2}{8} - \frac{1348}{16} = 100.00.$$

$$SS(AC) = \frac{(181 + 200 + 163 + 140)^2}{8} + \frac{(129 + 144 + 187 + 204)^2}{8} - \frac{1348^2}{16} = 25.00.$$

$$SS(BC) = \frac{(181 + 144 + 187 + 140)^2}{8} + \frac{(129 + 200 + 163 + 204)^2}{8} - \frac{1348^2}{16} = 121.00.$$

$$SS(ABC) = \frac{(129 + 200 + 187 + 140)^2}{8} + \frac{(181 + 144 + 163 + 204)^2}{8} - \frac{1348^2}{16} = 81.00$$

$$SSE = 3417 - (100 + 49 + 2401 + 100 + 25 + 121 + 81) = 540.00.$$

The ANOVA results are given in Table 9.15. Only the main factor C is significant at $\alpha = 0.05$.

TABLE 9.15

ANOVA for Example 9.8

Source of Variation	Sum of Squares	Degrees of Freedom	Mean Square	Computed F	F from Table A.3 ($\alpha = 0.05$)
A	100	1	100	1.48	5.32
B	49	1	49	<1	
C	2401	1	2401	35.57[a]	
AB	100	1	100	1.48	
AC	25	1	25	<1	
BC	121	1	121	1.79	
ABC	81	1	81	1.2	
Error	540	8	67.5	—	
Total	3417	15	—	—	

[a] Indicates significance at $\alpha = 0.05$.

TABLE 9.16

Contrasts and Effects for Example 9.8

	A	B	C	AB	AC	BC	ABC	Obs. 1	Obs. 2	Total
	−	−	−	+	+	+	−	97	84	181
	−	−	+	+	−	−	+	63	66	129
	−	+	−	−	+	−	+	96	104	200
	−	+	+	−	−	+	−	72	72	144
	+	−	−	−	−	+	+	87	100	187
	+	−	+	−	+	−	−	75	88	163
	+	+	−	+	−	−	−	104	100	204
	+	+	+	+	+	+	+	59	81	140
Contrast	40	28	−196	−40	20	−44	−36	—	—	—
Effect	5.0	3.5	−24.5	−5.0	2.5	−5.5	−4.5	—	—	—

Let us now examine the formulas for computing the sums of squares. There are k factors, and the number of replications is n. Then, there are $2^k \times n = m$ observations. The number of observations at each level of each of the n factors $= m/2$. Consider any main factor or interaction. Let T_1 be the sum of the observations at level "+" and T_2 be the sum of the observations at level "−". Then, the grand sum of all the m observations is $T = T_1 + T_2$. The sum of squares for this factor is computed as:

$$\text{Sum of squares} = (T_1^2 + T_2^2)/(m/2) - T^2/m$$
$$= (T_1 - T_2)^2/m$$

The quantity within the parentheses is called the *contrast* of this factor:

$$\text{Sum of squares for any factor or interaction} = \text{Contrast}^2/m \quad (9.69)$$

$$\text{Effect of any main factor or interaction} = \text{Contrast}/(m/2) \quad (9.70)$$

Table 9.16 contains all the contrasts and effects in Example 9.8.

9.6.2 Polynomial Equation

In general, the polynomial equation in a 2^3 full factorial experiment is

$$\hat{y} = \alpha'_0 + \alpha'_1 x_1 + \alpha'_2 x_2 + \alpha'_3 x_3 + \alpha'_{12} x_1 x_2 + \alpha'_{13} x_1 x_3 + \alpha'_{23} x_2 x_3 + \alpha'_{123} x_1 x_2 x_3$$

$$(9.71)$$

The orthogonal polynomial equation is

$$\hat{y} = \alpha_0 + \alpha X_1 + \alpha_2 X_2 + \alpha_3 X_3 + \alpha_{12} X_1 X_2 + \alpha'_{13} X_1 X_3 + \alpha_{23} X_2 X_3 + \alpha_{123} X_1 X_2 X_3$$

$$(9.72)$$

where

$$X_i = \left(\frac{x_i - t_i}{d_i}\right)$$

$$(9.73)$$

where $t_i = (\text{Max. level}_i + \text{Min. level}_i)/2$, and $d_i = (\text{Max. level}_i - t_i)$. The "intercept" α_0 is the grand average of all the observations, and α_i, α_{ij}, or α_{ijk} is equal to the associated effect/2.

It can be seen that X_i is a coded observation taking value -1 or $+1$. In Example 9.8,

i	t_i	d_i
1	$1\frac{1}{4}$	$\frac{1}{4}$
2	$1\frac{3}{16}$	$\frac{3}{16}$
3	$1\frac{1}{4}$	$\frac{1}{4}$

and the polynomial equation is

$$\hat{y} = 84.25 + 2.5\left(\frac{x_1 - 1.25}{0.25}\right) + 1.75\left(\frac{x_2 - 1.1875}{0.1875}\right) - 12.25\left(\frac{x_3 - 1.25}{0.25}\right)$$

$$- 2.5\left(\frac{x_1 - 1.25}{0.25}\right)\left(\frac{x_2 - 1.1875}{0.1875}\right) + 1.25\left(\frac{x_1 - 1.25}{0.25}\right)\left(\frac{x_3 - 1.25}{0.25}\right)$$

$$- 2.75\left(\frac{x_2 - 1.1875}{0.1875}\right)\left(\frac{x_3 - 1.25}{0.25}\right)$$

$$- 2.25\left(\frac{x_1 - 1.25}{0.25}\right)\left(\frac{x_2 - 1.1875}{0.1875}\right)\left(\frac{x_3 - 1.25}{0.25}\right)$$

$$(9.74)$$

where $x_1 = 1$ or $1\frac{1}{2}$, $x_2 = 1$ or $1\frac{3}{8}$, and $x_3 = 1$ or $1\frac{1}{2}$.

Residual and other diagnostic analyses must be performed before using the equation for predicting the response variable within the ranges of the main factors included in the experiment.[4] The linear model, analyses, and methodology used in developing the equation can be easily extended to more than three factors.

9.6.3 Factorial Experiments in Incomplete Blocks

Suppose that a trial is to be conducted using a 2^3 factorial design. To make the eight runs under conditions as homogeneous as possible, it is desirable that batches of raw material sufficient for the complete experiment for one replication be blended together. Suppose, however, that the available blender is only large enough to accommodate material for four runs. This means that blends mixed at two different times will have to be used. Hence, all the eight combinations will not have "identical" blends. In this case, the 2^3 design should be arranged in two blocks of four runs each to neutralize the effect of possible blend differences. The disadvantage with such an experimental setup is that certain effects are completely confounded or mixed with the blocks as a result of blocking. The number of effects confounded depends on the number of blocks required.

9.6.3.1 Experiments with Two Blocks

One effect is confounded in an experiment with two blocks. Usually the highest order interaction is selected to be confounded. For example, if the three-factor interaction effect is confounded in a 2^3 design with two blocks, then only the main effects and two-factor interactions can be studied in this experiment. The method of distribution of treatment combinations between the blocks is illustrated in the next example.

Example 9.9

Construct two incomplete blocks of a 2^3 design:

1. Define the effect to be confounded, called the *defining contrast*. In Example 9.9, the logical defining contrast is the three-factor interaction, ABC.
2. Write all the 2^k combinations in a table with "−" representing low levels and "+" representing high levels. The combinations for Example 9.9 are given in Table 9.17.

TABLE 9.17

Design Table for Example 9.9

A	B	C	ABC
−	−	−	−
−	−	+	+
−	+	−	+
−	+	+	−
+	−	−	+
+	−	+	−
+	+	−	−
+	+	+	+

3. All the combinations that have the sign "–" in column ABC in Table 9.17 belong to one block, whereas the other combinations form the second block:

Block 1 "+"			Block 2 "–"		
A	B	C	A	B	C
–	–	+	–	–	–
–	+	–	–	+	+
+	–	–	+	–	+
+	+	+	+	+	–

If this is to be the experiment that blends two different batches because of the capacity of the blender, then the combinations in block 1 will have to be made out of blend 1, and blend 2 should supply the material for the combinations in block 2.

Example 9.10

Divide the combinations in a 2^4 factorial experiment into two blocks, using ACD as the defining contrast (see Table 9.18).

9.6.3.2 Experiments with Four Incomplete Blocks

If the treatment combinations of a 2^k factorial experiment is to be divided into four incomplete blocks, then the experimenter chooses any two defining contrasts (those effects that will be confounded with blocks). A third effect, called the *generalized interaction* of the two defining contrasts, is automatically confounded with the blocks. Thus, a total of three effects will be confounded with blocks in an experiment with four incomplete blocks.

TABLE 9.18

Design Table for Example 9.10

A	B	C	D	ACD	Block 1 ("–" in ACD)	Block 2 ("+" in ACD)
–	–	–	–	–	*	
–	–	–	+	+		*
–	–	+	–	+		*
–	–	+	+	–	*	
–	+	–	–	–	*	
–	+	–	+	+		*
–	+	+	–	+		*
–	+	+	+	–	*	
+	–	–	–	+		*
+	–	–	+	–	*	
+	–	+	–	–	*	
+	–	+	+	+		*
+	+	–	–	+		*
+	+	–	+	–	*	
+	+	+	–	–	*	
+	+	+	+	+		*

Example 9.11

Divide a 2^4 factorial experiment into four incomplete blocks:

1. The experimenter should choose two defining contrasts—two effects that are to be confounded. Suppose that, in our example, the experimenter chooses AB and CD as the defining contrasts.

2. Determine the third effect (generalized interaction) that will be confounded by multiplying both the defining contrasts and choosing the letters with odd exponents only. In this example, $AB \times CD = ABCD$ is the generalized interaction, because each of the letters A, B, C, and D has an exponent of one. More examples of 2^4 factorials are given below:

Defining Contrasts		Generalized Interaction
AB	ABC	C
ABD	ABC	CD
BCD	AB	ACD

3. Group the treatment combinations into four blocks based on the signs in the defining contrasts selected. In this example, the blocks are as follows (see Table 9.19):

AB	CD	Block
−	−	1
−	+	2
+	−	3
+	+	4

4. The observations corresponding to the treatment combinations in each block should be collected under identical conditions.

9.6.4 Fractional Factorial Experiments

The 2^k factorial experiment can become quite large when k is large. In many experimental situations, it can be assumed that certain higher order interactions are negligible or, even if they are not negligible, the experimenter may not be interested in them and thus it would be a waste of experimental effort to use the complete factorial experiment. When k is large, the experimenter can make use of a fractional factorial experiment in which only one half, one fourth, or even one eighth of the total factorial design is actually carried out.

TABLE 9.19

Design Table for Example 9.11

A	B	C	D	AB	CD	Block 1	Block 2	Block 3	Block 4
−	−	−	−	+	+				*
−	−	−	+	+	−			*	
−	−	+	−	+	−			*	
−	−	+	+	+	+				*
−	+	−	−	−	+		*		
−	+	−	+	−	−	*			
−	+	+	−	−	−	*			
−	+	+	+	−	+		*		
+	−	−	−	−	+		*		
+	−	−	+	−	−	*			
+	−	+	−	−	−	*			
+	−	+	+	−	+		*		
+	+	−	−	+	+				*
+	+	−	+	+	−			*	
+	+	+	−	+	−			*	
+	+	+	+	+	+				*

9.6.4.1 One Half Fractional Factorial Design (Half-Replicate)

The construction of a half-replicate design is identical to the allocation of a 2^k factorial experiment into two blocks. First a defining contrast is selected that is to be confounded, then the two blocks are constructed and either of them is selected as the design to be carried out.

As an example, consider a 2^4 factorial experiment in which we wish to use a half-replicate. The defining contrast $ABCD$ is chosen and two blocks are as follows:

Block 1 ("+" for $ABCD$)				Block 2 ("−" for $ABCD$)			
A	B	C	D	A	B	C	D
−	−	−	−	−	−	−	+
−	−	+	+	−	−	+	−
−	+	−	+	−	+	−	−
−	+	+	−	−	+	+	+
+	−	−	+	+	−	−	−
+	−	+	−	+	−	+	+
+	+	−	−	+	+	−	+
+	+	+	+	+	+	+	−

Either block can be selected. Let us assume that the experimenter selects block 1 and collects data for the eight combinations in that block. Table 9.20 contains these eight combinations in the block 1 experiment, with all possible main factors and interactions in a 2^4 full factorial experiment. Even though having two or more replications enables us to compute an explicit sum of squares for error, it does not increase the number of sums of squares due to main factors or interactions. The number of sums of squares that can be computed (other than SST) using the above data is $8 - 1 = 7$. The total number of possible effects (main factors and their interactions) in a 2^4 experiment is

TABLE 9.20

Treatment Combinations in Block 1

A	B	C	D	AB	AC	AD	BC	BD	CD	ABC	ABD	ACD	BCD	ABCD
−	−	−	−	+	+	+	+	+	+	−	−	−	−	+
−	−	+	+	+	−	−	−	−	+	+	+	−	−	+
−	+	−	+	−	+	−	−	+	−	+	−	+	−	+
−	+	+	−	−	−	+	+	−	−	−	+	+	−	+
+	−	−	+	−	−	+	+	−	−	+	−	−	+	+
+	−	+	−	−	+	−	−	+	−	+	−	+	+	+
+	+	−	−	+	−	−	−	−	+	−	−	+	+	+
+	+	+	+	+	+	+	+	+	+	+	+	+	+	+

15, out of which interaction $ABCD$ is not "present" in block 1, because all the combinations in this block have the same sign, "+". This leaves out 14 effects that are present in the experiment, which means that each of the seven sums of squares is shared by two effects. It can be seen from Table 9.20 that there are seven pairs of effects (main factors and interactions) such that the effects in each pair have the same sequence of "−" and "+" signs and the same sum of squares. Examples of such pairs are A and BCD, B and ACD, and so on. The effects in a pair are called *aliases*. The aliases in each group (pair) can be obtained by deleting the letters with an even exponent from the product of the effect (main factor or interaction) and the defining contrast. For example, the alias of A is $A \times ABCD = A^2BCD = BCD$.

The aliases in this one half fractional factorial design are

$$A + BCD$$
$$B + ACD$$
$$C + ABD$$
$$D + ABC$$
$$AB + CD$$
$$AC + BD$$
$$AD + BC$$

In summary, in a one half fractional factorial design, the sum of squares of the defining contrast cannot be computed. In addition, there are exactly two effects (main factors and or interactions) in each alias group. If the test statistic obtained from the sum of squares of an alias group is significant, we cannot determine which one of the members of that group is the cause of significance without supplementary statistical evidence. However, fractional factorial designs have their greatest use when k is quite large and there is some previous knowledge concerning the interactions. It becomes evident that one should always be aware of what the alias structure is for a fractional experiment before finally adopting the experimental plan. Proper choice of defining contrasts and awareness of the alias structure are important considerations before an experimental design is selected. The calculation of the sums of squares is done in the same way as before, keeping in mind the above limitations.

9.6.4.2 *One Quarter Fractional Factorial Design (Quarter-Replicate)*

The construction of a quarter-replicate design is identical to the allocation of a 2^k factorial experiment into four blocks. Two defining contrasts are specified to partition the 2^k combinations into four blocks. Any one of the four blocks can be selected for performing the experiment and analysis. In this design, the defining contrasts and the generalized interaction are not "present" because each of these will have the same sign ("−" or "+") in any block selected.

Let us consider a one quarter fractional design of a 2^5 factorial, constructed using *ABD* and *ACE* as the defining contrasts. The generalized interaction is *BCDE*. In this design, *ABD*, *ACE*, and *BCDE* are not "present" because each of these will have the same sign ("−" or "+") in any of the four blocks. This leaves out $2^5 - 1 - 3 = 28$ effects (main factors and their interactions) that are present in this design. As the total number of treatment combinations in this design is $^1/_4(2^5) = 8$, only seven $(8 - 1)$ sums of squares can be computed. This means that each sum of squares is shared by $28/7 = 4$ effects (main factors/interactions), hence there are four aliases in each group. The aliases in each group can be obtained by deleting the letters with even exponents from the products of any one effect (main factor or interaction) and each defining contrast and the generalized interaction. For example, the aliases of *A* are

$$A \times ABD = A^2BD = BD$$
$$A \times ACE = A^2CE = CE$$
$$A \times BCDE = ABCDE$$

This means that *A*, *BD*, *CE*, and *ABCDE* share the same sum of squares, mean squares and test statistics.

The following formulas are applicable to incomplete block designs and fractional designs of a 2^k factorial experiment.

1. Full factorial:

    Number of factors $= k$.

    Total number of combinations in a full factorial experiment $= 2^k$.

    Total number of effects (main factors/interactions) present in a 2^k full factorial experiment $= 2^k - 1$.

2. Incomplete blocks:

    Number of blocks $= 2^q$, $q = 1, 2, ..., k - 1$.

    Number of combinations in each block $= 2^k/2^q = 2^{k-q}$.

    Number of defining contrasts needed to generate 2^q blocks $= q$.

    Total number of effects confounded within blocks $= 2^q - 1$.

    Total number of generalized interactions $= 2^q - 1 - q$.

3. Fractional factorial:

    Fraction $= 1/2^q$.

    Notation $= 2^{k-q}$.

    Number of treatment combinations in the fraction $= 2^{k-q}$.

Number of sums of squares that can be computed = $2^{k-q} - 1$.

Number of effects (due to main factors/interactions) that are present in the design = $2^k - 2^q$.

Number of alias groups (one sum of squares per each alias group) = $2^{k-q} - 1$.

Number of aliases in each group = 2^q.

9.6.4.3 Construction of Fractional Factorial Designs

The type of alias relationship present in a fractional factorial design is defined by its resolution.

1. *Resolution III designs*: In this type of designs,
   - No main factor is aliased with any other main factor.
   - Main factors are aliased with two-factor interactions.
   - Two-factor interactions are aliased with other two-factor interactions.

   Examples include 2^{3-1} and 2^{5-2} designs.

2. *Resolution IV designs*: These are the designs in which:
   - No main factor is aliased with either another main factor or a two-factor interaction.
   - Two-factor interactions are aliased with other two-factor interactions.

   Examples include 2^{4-1} and 2^{6-2} designs.

3. *Resolution V designs*: In these designs,
   - No main factor is aliased with either another main factor or a two-factor interaction.
   - No two-factor is aliased with other two-factor interactions.
   - Two-factor interactions are aliased with three-factor interactions.

   Examples include 2^{5-1} and 2^{6-1} designs.

Table 9.21 contains recommended defining contrasts for selected fractional factorial designs and the resulting resolutions. For more designs, please refer to Montgomery.[3]

A basic design is a 2^a full factorial design where $a = k - q$. For example, the basic design of a 2^{7-3} fractional factorial design is a 2^4 full factorial design. The number of rows (treatment combinations) in a 2^{k-q} fractional factorial design is equal to the number of rows (treatment combinations) in the associated basic design.

9.6.4.4 Steps in the Construction of a Fractional Factorial Design

1. Based on the number of factors to be tested (included in the experiment) and the resolution desired and/or the maximum number of feasible experiments, select a fractional factorial design from Table 9.21.

TABLE 9.21

Recommended Defining Contrasts for Selected Fractional Factorial Designs

Number of Factors (k)	Fractional Design (2^{k-q})	Resolution	Number of Experiments/Treatment Combinations	Defining Contrasts
3	2^{3-1} (1/2)	III	4	ABC
4	2^{4-1} (1/2)	IV	8	ABCD
5	2^{5-2} (1/4)	III	8	ABD,ACE
	2^{5-1} (1/2)	V	16	ABCDE
6	2^{6-3} (1/8)	III	8	ABD, ACE, BCF
	2^{6-2} (1/4)	IV	16	ABCE, BCDF
7	2^{7-4} (1/16)	III	8	ABD, ACE, BCF, ABCG
	2^{7-3} (1/8)	IV	16	ABCE, BCDF, ACDG
	2^{7-2} (1/4)	IV	32	ABCDF, ABDEG
8	2^{8-4} (1/16)	IV	16	BCDE, ACDF, ABCG, ABDH
	2^{8-3} (1/8)	IV	32	ABCF, ABDG, BCDEH
9	2^{9-5} (1/32)	III	16	ABCE, BCDF, ACDG, ABDH, ABCDJ
	2^{9-4} (1/16)	IV	32	BCDEF, ACDEG, ABDEH, ABCEJ
	2^{9-3} (1/8)	IV	64	ABCDG, ACEFH, CDEFJ

Example 9.12

An experiment is to be conducted to test the effect of seven factors on some response variable. The experimenter is satisfied with Resolution III. It is a 2^{7-4} fractional design.

2. Select the defining contrast(s) from Table 9.21. From Table 9.21, the recommended defining contrasts for Example 9.12 are *ABD*, *ACE*, *BCF*, and *ABCG*.

3. Start with the basic design, which is a 2^a full factorial design where $a = k - q$. Table 9.21 contains the signs for the first $a = k - q$ factors in the fractional design. In Example 9.12, the basic design is 2^3 full factorial, given in Table 9.22, which contains signs for factors *A*, *B*, and *C* in the 2^{7-4} fractional design.

4. Using the alias relationship, identify the columns for the remaining q factors. In Example 9.12,

   One alias of *D* is $D \times ABD = AB$ (*D* and *AB* share the same column).

   One alias of *E* is $E \times ACE = AC$ (*E* and *ACE* share the same column).

   One alias of *F* is $F \times BCF = BC$ (*F* and *BCF* share the same column).

   One alias of *G* is $G \times ABCG = ABC$ (*G* and *ABC* share the same column).

TABLE 9.22

Basic Design for Example 9.12

A	B	C	AB	AC	BC	ABC
−	−	−	+	+	+	−
−	−	+	+	−	−	+
−	+	−	−	+	−	+
−	+	+	−	−	+	−
+	−	−	−	−	+	+
+	−	+	−	+	−	−
+	+	−	+	−	−	−
+	+	+	+	+	+	+

TABLE 9.23

2^{7-4} Fractional Factorial Design for Example 9.12

A	B	C	D	E	F	G
−	−	−	+	+	+	−
−	−	+	+	−	−	+
−	+	−	−	+	−	+
−	+	+	−	−	+	−
+	−	−	−	−	+	+
+	−	+	−	+	−	−
+	+	−	+	−	−	−
+	+	+	+	+	+	+

Table 9.22 is modified by replacing column headings $AB, AC, BC,$ and ABC by $D, E, F,$ and G, respectively, resulting in the final 2^{7-4} fractional factorial design, given in Table 9.23. In real-life applications, signs "+" and "−" in the design table are replaced by the actual levels of the factors.

Example 9.13

Table 9.24 contains the log sheet of an experiment conducted to test the effects of seven factors on the hardness of a powdered metal component:

A	Material composition	5% (−) and 10% (+)
B	Binder type	1 (−) and 2 (+)
C	Position in the basket	Bottom (−) and top (+)
D	Heat treatment temperature	800°F (−) and 900°F (+)
E	Quenching bath medium	Water (−) and oil (+)
F	Annealing temperature	300°F (−) and 400°F (+)
G	Speed of conveyor belt in annealing oven	2 feet/minute (−) and 4 feet/minute (+)

TABLE 9.24

Log Sheet of the 2^{7-4} Fractional Factorial Experiment in Example 9.13

A Material Composition (%)	B Binder Type	C Position in Basket	D Heat Treatment Temperature (°F)	E Quenching Bath	F Annealing Temperature (°F)	G Speed of Conveyor (fpm)
5	1	Bottom	900	Oil	400	2
5	1	Top	900	Water	300	4
5	2	Bottom	800	Oil	300	4
5	2	Top	800	Water	400	2
10	1	Bottom	800	Water	400	4
10	1	Top	800	Oil	300	2
10	2	Bottom	900	Water	300	2
10	2	Top	900	Oil	400	4

TABLE 9.25

Data for Example 9.14

A	B	C	D	E	Replication 1	Replication 2	Total
−	−	−	+	+	71	72	143
−	−	+	+	−	106	100	206
−	+	−	−	+	59	62	121
−	+	+	−	−	91	94	185
+	−	−	−	−	122	119	241
+	−	+	−	+	91	94	185
+	+	−	+	−	131	119	250
+	+	+	+	+	85	69	154

Example 9.14

One quarter of a 2^5 factorial experiment (2^{5-2}) was conducted to test the effects of the following five factors on the descent time of a paper helicopter:

A	Wing length	1" (−) and 1½" (+)
B	Body width	1" (−) and 1⅜" (+)
C	Tail length	1" (−) and 1½" (+)
D	Paper type	1 (−) and 2 (+)
E	Ballast	Without (−) and with (+)

The defining contrasts are *ABD* and *ACE* with Resolution III. Table 9.25 contains the observations collected in the experiment, which was replicated twice. The generalized interaction is *BCDE*. There are seven alias groups with four effects (main factors and interactions) in each group. Table 9.26 contains the ANOVA for this problem.

TABLE 9.26

ANOVA for Example 9.14

Source of Variation	Sum of Squares	Degrees of Freedom	Mean Square	Computed F	F from Table A.3 ($\alpha = 0.05$)
$A + BD + CE$ $+ ABCDE$	1914.06	1	1914.06	64.86[a]	5.32
$B + AD + CDE$ $+ ABCE$	264.06	1	264.06	8.95[a]	
$C + AE + BDE$ $+ ABCD$	39.06	1	39.06	1.32	
$D + AB + BCE$ $+ ACDE$	27.56	1	27.56	<1	
$E + AC + BCD$ $+ ABDE$	4865.06	1	4865.06	164.8[a]	
$BC + DE + ABE$ $+ ACD$	95.06	1	95.06	3.22	
$BE + CD + ABC$ $+ ADE$	105.06	1	105.06	3.56	
Error	236.05	8	29.51		
Total	7546.44	15			

[a] Indicates significance at $\alpha = 0.05$.

9.6.4.5 Conclusions

The main effects A, B, and E and their aliases are significant at $\alpha = 0.05$. In the absence of additional evidence, it is not possible to ascertain which of the effects in a particular alias group is (are) significant. Usually, the main effects are presumed to cause significance in an alias group.

Equations relating the significant main factors and interactions to the response variables can be developed using regression methods or the polynomial equation method, described earlier in this chapter. Assuming that A, B, and E are the only significant factors, the following polynomial equation is developed for Example 9.14:

$$\hat{y} = 92.81 + 10.94 \left(\frac{x_1 + 1.25}{0.25} \right) - 4.06 \left(\frac{x_2 + 1.1875}{0.1875} \right) - 17.44 X_5, \quad (9.75)$$

where $X_5 = -1$ when used without ballast and 1 when used with ballast.

After developing appropriate equations, the levels of the factors that optimize the response variable can be determined using response methodology or the desirability function approach.[6]

9.6.5 Addition of Center Points to the 2^k Design

A major problem in using a 2^k factorial design is the assumption of linearity in the effect of signal factors on the response variables. An experiment with three levels for the factors suspected to have a nonlinear effect on the response variable (for example, 3^k factorial experiment) can be used to remedy this problem. Adding a center point to a 2^k factorial experiment can also take care of this problem with a 2^k factorial experiment. Replicating the center point also provides an independent estimate of error.

Example 9.15

Consider a 2^2 factorial design in our paper helicopter design problem. Let the two signal factors be the wing length (A) and body width (B). The following are the data for one replicate:

	Wing Length (A)	
Body Width (B)	1" (−)	1½" (+)
1" (−)	63	87
1⅜" (+)	72	104

If we want to test for the nonlinear effect of the factors on the descent time, then we can add a center point to this design as shown in Figure 9.8. Let us assume that the center point is replicated four times and the readings are 55, 56, 58, and 58. As the center point is replicated four times, the MSE is computed using only the four observations at this point:

$$\text{MSE} = \text{SSE}/(n_c - 1) \tag{9.76}$$

where n_c is the number of replications at the center point, and

$$\text{SSE} = \sum_{i=1}^{n_c} (y_i - \bar{y}_c)^2 \tag{9.77}$$

where $\bar{y}_c$ is mean of replications at the center point.

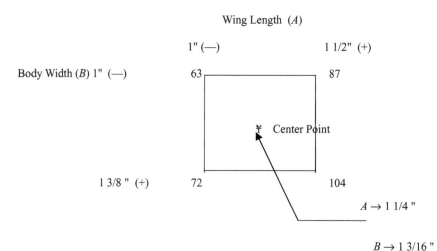

FIGURE 9.8
Addition of center point.

TABLE 9.27

ANOVA for Example 9.15

Source of Variation	Sum of Squares	Degrees of Freedom	Mean Square	Computed F	F from Table A.3 ($\alpha = 0.05$)
A	784	1	784	348.44[a]	10.13
B	169	1	169	75.11[a]	
AB	16	1	16	7.11	
Curvature	1225.13	1	1225.13	555.56[a]	
Error	6.75	4 − 1 = 3	2.25		
Total	2200.88	7			

[a] indicates significance at $\alpha = 0.05$.

In Example 9.15, $\bar{y}_c = (55 + 56 + 58 + 58)/4 = 56.75$, SSE $= (55 - 56.75)^2 + (56 - 56.75)^2 + (58 - 56.75)^2 + (58 - 56.75)^2 = 6.75$, and MSE $= 6.75/(4-1) = 2.25$. The sum of squares due to curvature is computed as:

$$SS_{CURVATURE} = \frac{n_f n_c (\bar{y}_f - \bar{y}_c)^2}{(n_f + n_c)} \qquad (9.78)$$

where $\bar{y}_f$ is the average of all observations, excluding replications at the center point, and n_f is the total number of observations, excluding replications at the center point. In Example 9.15, $\bar{y}_f = (63 + 87 + 72 + 104)/4 = 81.5$, and

$$SS_{CURVATURE} = \frac{4 \times 4 \times (81.5 - 56.75)^2}{(4+4)} = 1225.13$$

The other sums of squares (SST, SSA, SSB, SSAB, and SSE) are computed as in any 2^k design. The ANOVA results are given in Table 9.27.

It can be seen that both the main factors and the curvature effect are significant at $\alpha = 0.05$. A more detailed experiment with more levels of A and B is required to capture the potential nonlinear effects of A and B on the response variable.

9.7 Taguchi's Orthogonal Arrays

Taguchi simplified the procedure used in the design of fractional factorial experiments by constructing orthogonal arrays (refer to Taguchi[7] and Phadke[5] for details of the arrays). His orthogonal arrays are fractional factorial designs of two levels, three levels, and mixtures of two, three, four, and five levels. The arrays are denoted by L_a, where a is the number of rows in the

TABLE 9.28

2^3 Basic Design

−	−	−	+	+	+	−
−	−	+	+	+	−	−	+
−	+	−	−	+	−	+
−	+	+	−	−	+	−
+	−	−	−	−	+	+
+	−	+	−	+	−	−
+	+	−	+	−	−	−
+	+	+	+	+	+	+

array and indicates the number of experiments that need to be performed for that array. For example, an L_8 array has eight rows and L_9 has nine rows. The L_8 array can be used as a 2^{k-q} fractional factorial design for any k and q such that, $k - q = 3$. From Section 9.6.4, we know that the basic design of a 2^{k-q} fractional factorial design is a 2^{k-q} full factorial design. Hence, the L_8 array can be considered a basic design for all 2^{k-q} fractional factorial designs, with $k - q = 3$. Consider the 2^3 full factorial design given in Table 9.28, which has no column headings. Each column is assigned to a factor. Without column headings, we can assign any column to any factor, keeping in mind the columns that contain the interactions of other columns. This is important in fractional factorial designs, because of the presence of alias groups. Taguchi addressed the interactions in tables and graphs.[7] His L_8 design can be obtained from the basic design in Table 9.28 by reversing the order of rows; replacing "+" and "−" by 1 and 2, respectively, and interchanging columns 3 and 4. Similarly, the L_{16}, L_{32}, and L_{64} arrays are basic designs for all 2^{k-q} fractional factorial designs with k and q such that $k - q = 4, 5$, and 6, respectively.

Taguchi developed other types of orthogonal arrays that are difficult to design using classical design of experiments methodology. For example the L_{18} array has one column with two levels and seven columns with three levels. The advantage in using these arrays is that the experimenter can test several factors with unequal levels using very few experiments. The main disadvantage in such arrays is that it is impossible to test the effects of any interaction. The only option the experimenter has is to assume that all the interaction effects are negligible, which may not be true in many real-life situations.

9.8 Design of Experiments Methodology

Barton[1] presents an excellent discussion on the methodology to be used in the design of experiments; the flow chart in Figure 9.9 is a summary of the steps to be used.

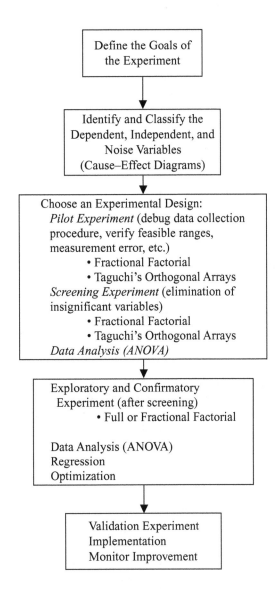

FIGURE 9.9
Design of experiments methodology.

9.9 References

1. Barton, R.R., *Graphical Methods for the Design of Experiments*, Lecture Notes in Statistics, Springer-Verlag, New York, 1999.
2. Box, G.E.P. and Cox, D.R., An analysis of transformations, *J. Roy. Stat. Soc. B*, 26 (1964).

3. Montgomery, D.C., *Design of Experiments,* 3rd ed., John Wiley & Sons, New York, 1997.
4. Neter, J., Kutner, M.H., Nachtsheim, C.J., and Wasserman, W., *Applied Linear Statistical Methods,* 3rd ed., Irwin, Chicago, IL, 1996.
5. Phadke, M.S., *Quality Engineering Using Robust Design,* Prentice Hall, New York, 1989.
6. Wu, C.F.J. and Hamada, M., *Experiments: Planning, Analysis, and Parameter Design Optimization,* John Wiley & Sons, New York, 2000.
7. Taguchi, G., *System of Experimental Design,* Vols. 1 and 2, UNIPUB/Kraus International Publications, New York, and American Supplier Institute, Dearborn, MI, 1987.

9.10 Problems

1. A manufacturer in the steel industry wants to test the effect of the method of manufacture on the tensile strength of a particular type of steel. Four different methods have been tested, and the data shown in Table 9.29 have been collected. Test the significance of the effect at $\alpha = 0.05$.

2. Data have been collected to test whether the thickness of a polysilicon coating depends upon the deposition temperature (see Table 9.30). Test whether there is any difference in the thickness at different temperature levels at $\alpha = 0.05$.

3. Prove that $\sum_{i=1}^{t}\sum_{j=1}^{r}(\bar{y}_{i\cdot} - \bar{y}_{\cdot\cdot})(y_{ij} - \bar{y}_{i\cdot}) = 0$.

TABLE 9.29

Data for Problem 1

Method	Tensile Strength			
1	2650	2765	2750	2600
2	2985	2975	2865	2890
3	2775	2620	2690	2700
4	2900	2885	2850	2950

TABLE 9.30

Data for Problem 2

Deposition Temperature	Polysilicon Thickness (Coded Data)			
Level 1	110	112	109	115
Level 2	105	99	100	103
Level 3	115	120	109	117
Level 4	102	110	108	102

TABLE 9.31

Data for Problem 4

	Operators			
	1	2	3	4
Output	90	130	125	150
	100	145	140	145
	95	120	135	135
	80	135	140	160

TABLE 9.32

Data for Problem 5

Batches				
1	2	3	4	5
8.5	8.4	8.4	8.5	8.9
8.6	8.5	8.9	8.4	8.6
8.2	8.3	8.2	8.2	8.2

TABLE 9.33

Data for Problem 6

	Factor A					
Factor B	600		650		700	
1	142	113	115	129	135	120
	126	130	109	98	115	110
2	122	104	112	104	144	132
	118	138	100	119	120	139
3	110	126	122	100	142	130
	138	120	118	109	110	125

4. The management wants to test whether there is any difference in the output of operators on the shop floor. Four operators were selected at random from the shop floor, and the data in Table 9.31 represent the output of the operators. Conduct an ANOVA at $\alpha = 0.01$. Estimate the variance due to the operators.

5. A chemical engineer has reason to believe that the batches of raw material used in the manufacture of a compound differ significantly in the nitrogen content. Currently, the company uses many batches; for his study, the engineer randomly selected five of these batches and recorded three observations on each batch. The data in Table 9.32 were obtained. Conduct an ANOVA at $\alpha = 0.01$. Estimate the variance due to batches.

6. A two-factor experiment was conducted to test whether the given factors affect the response variable. The data shown in Table 9.33 were collected for three levels of A and three levels of B with four replicates.

   a. Using $\alpha = 0.05$, test for the significance of all possible effects.

   b. Estimate the percentage contributions of all possible effects and the error.

7. An engineer suspects that the cutting speed (in inches) and the feed rate (inches/minute) influence the surface finish of a metal part. For this study, the engineer collected four cutting speeds and three feed rates and conducted a two-factor experiment to obtain the data shown in Table 9.34 with three replications.

   a. Conduct an ANOVA at $\alpha = 0.05$ to test the significance of all possible effects.

   b. Estimate the percentage contributions of all the possible effects and the error.

8. An experiment was performed to assess the effect of engine oil on the life of lawn mower engines. The data shown in Table 9.35 (engine life in hours which are coded) was collected. There are two replications. Conduct an ANOVA to test all possible effects in this experiment. Use $\alpha = 0.01$.

9. The effects of cutting speed and feed on the surface finish of a component were investigated in a two-factor experiment. Each factor was set at two levels and two observations were collected for each of the four combinations. The coded observations are shown in Table 9.36.

   a. Conduct an ANOVA at $\alpha = 0.05$ and test for all possible effects.

   b. Calculate the percentage contribution of all the effects.

TABLE 9.34

Data for Problem 7

Cutting Speed (RPM)	Feed Rate (inches/minute)								
	0.10			0.20			0.30		
100	54	60	65	79	68	80	99	88	94
200	98	89	94	98	106	90	100	98	108
300	109	98	89	100	108	94	99	82	102
400	112	100	92	94	104	108	98	116	106

TABLE 9.35

Data for Problem 8

Oil Type 1 Mower Make		Oil Type 2 Mower Make	
I	II	III	IV
1	2	5	8
2	3	6	9

TABLE 9.36

Data for Problem 9

Feed (B)	Cutting Speed (A) (RPM)	
	200	300
2	2 3	4 5
4	6 7	1 2

TABLE 9.37

Data for Problem 10

A	B	C	D	Replication 1	Replication 2
−1	−1	−1	−1	200	197
−1	−1	−1	1	194	189
−1	−1	1	−1	201	202
−1	−1	1	1	193	196
−1	1	−1	−1	190	208
−1	1	−1	1	195	198
−1	1	1	−1	205	203
−1	1	1	1	199	197
1	−1	−1	−1	188	190
1	−1	−1	1	180	178
1	−1	1	−1	192	193
1	−1	1	1	205	204
1	1	−1	−1	199	200
1	1	−1	1	179	175
1	1	1	−1	187	185
1	1	1	1	185	186
Total				3092	3101

  c. Construct an orthogonal polynomial equation for the data.

  d. Using the equation in item c, predict the surface finish when cutting speed is 200 RPM and feed is 4; compute the residual.

10. In an experiment to test the effects of four factors—A, B, C, and D—on a response variable, a 2^4 full factorial experiment was conducted with two replications. The data are given in Table 9.37. Conduct an ANOVA at $\alpha = 0.05$.

11. A study was performed to determine the effect of material composition (A), heat treatment temperature (B), and annealing temperature (C) on the tensile strength achieved. Each replicate of the 2^3 design was run in two blocks, because only four components can be manufactured using the same batch of raw material. Two replicates were run, with ABC confounded in Replicates I and II. The coded data are as follows:

	Replicate I						
	Block 1				Block 2		
A	*B*	*C*		*A*	*B*	*C*	
−	−	−	−3	+	−	−	1
+	+	−	4	−	+	−	−1
+	−	+	3	−	−	+	−1
−	+	+	2	+	+	+	6

	Replicate II						
	Block 3				Block 4		
A	*B*	*C*		*A*	*B*	*C*	
−	−	−	−1	+	−	−	2
+	+	−	4	−	+	−	0
+	−	+	1	−	−	+	0
−	+	+	1	+	+	+	6

Conduct an ANOVA at $\alpha = 0.5$.

12. A one half fractional experiment was performed according to Table 9.38, which also contains the data. Conduct an ANOVA at $\alpha = 0.05$.

13. For the fractional factorial designs given below,

$$2^{5-1} \quad 2^{6-1} \quad 2^{6-2} \quad 2^{6-3} \quad 2^{7-2} \quad 2^{7-3} \quad 2^{7-4} \quad 2^{8-4}$$

indicate the following quantities:

a. Number of defining contrasts

b. Number of generalized interactions

c. Number of alias groups

d. Number of aliases in each group

14. Construct a 2^{6-2} fractional factorial design. List all the aliases.

15. Construct a 2^{7-3} fractional factorial design. List all the aliases.

16. The Research & Development division of a company wants to test the effect of three factors—*A*, *B*, and *C*—on a response variable. When designing this 2^3 factorial experiment without replication, the project team proposes using the design defined in Table 9.39 to accommodate the constraint that only two experiments can be performed using materials from the same blend (batch).

a. Briefly critique the design proposed by the team. More specifically, state the difficulties, if any, that might be encountered when interpreting the results. Do not criticize the constraint, as the team cannot do anything about that. Also, the team cannot replicate the experiment.

b. If you think that there are disadvantages in this design, recommend a better design that will remedy them, while at the same

TABLE 9.38

Data for Problem 12: 2^4 Factorial Experiment

A	B	C	D	AB	AC	AD	BC	BD	CD	ABC	ABD	ACD	BCD	ABCD	Replication 1	Replication 2	Total
−	−	−	−	+	+	+	+	+	+	−	−	−	−	+	5	6	11
+	−	−	−	−	−	−	+	+	+	+	+	+	−	−			
−	+	−	−	−	+	+	−	−	+	+	+	−	+	−	11	12	23
+	+	−	−	+	−	−	−	−	+	−	−	+	+	+			
−	−	+	−	+	−	+	−	+	−	+	−	+	+	−	32	30	62
+	−	+	−	−	+	−	−	+	−	−	+	−	+	+	15	18	33
−	+	+	−	−	−	+	+	−	−	−	+	+	−	+			
+	+	+	−	+	+	−	+	−	−	+	−	−	−	−	12	14	26
−	−	−	+	+	+	−	+	−	−	−	+	+	+	−			
+	−	−	+	−	−	+	+	−	−	+	−	−	+	+	16	12	28
−	+	−	+	−	+	−	−	+	−	+	−	+	−	+			
+	+	−	+	+	−	+	−	+	−	−	+	−	−	−	24	26	50
−	−	+	+	+	−	−	−	−	+	+	+	−	−	+	10	12	22
+	−	+	+	−	+	+	−	−	+	−	−	+	−	−			
−	+	+	+	−	−	−	+	+	+	−	−	−	+	−			
+	+	+	+	+	+	+	+	+	+	+	+	+	+	+			

TABLE 9.39

Data for Problem 16: Team's Design

A	B	C	AB	AC	BC	ABC	Shift	Blend
−	−	−	+	+	+	−	1	1
−	−	+	+	−	−	+	1	2
−	+	−	−	+	−	+	2	1
−	+	+	−	−	+	−	2	2
+	−	−	−	−	+	+	1	1
+	−	+	−	+	−	−	1	2
+	+	−	+	−	−	−	2	1
+	+	+	+	+	+	+	2	2

TABLE 9.40

Data for Problem 17[a]

	Cutting Speed (A)			
	−300 RPM		+350 RPM	
	Hardness (C)			
Feed Rate (B)	−30	+40	−30	+40
−0.002	10	15	8	13
+0.004	14	16	10	12

[a] Center point $(A \rightarrow 325$ RPM; $B \rightarrow 0.003$; $C \rightarrow 35)$.

time satisfying the given constraint. Please note that the team cannot replicate the experiment. Indicate your design by writing the shifts and blends in which each of the eight experiments is to be run in your design.

State the basis (or bases) of your design (that is, how you came up with your design) and the disadvantages of your design.

17. In a 2^3 full factorial experiment to test the effects of cutting speed (A), feed rate (B) and hardness of material (C) on the surface finish, one observation per each of the eight combinations was collected. In order to obtain an independent estimate of the error and to test the curvature effect of the factors on the response variable, the experimenter replicated the experiment four times at the center point. The coded observations are given in Table 9.40.

Conduct an ANOVA to test the effects of all possible effects at $\alpha = 0.05$.

Appendix

CONTENTS

TABLE A.1

Standard Normal Cumulative Probabilities

z	0.00	0.01	0.02	0.03	0.04	0.05	0.06	0.07	0.08	0.09
0.00	0.0000	0.0040	0.0080	0.0120	0.0160	0.0199	0.0239	0.0279	0.0319	0.0359
0.10	0.0398	0.0438	0.0478	0.0517	0.0557	0.0596	0.0636	0.0675	0.0714	0.0753
0.20	0.0793	0.0832	0.0871	0.0910	0.0948	0.0987	0.1026	0.1064	0.1103	0.1141
0.30	0.1179	0.1217	0.1255	0.1293	0.1331	0.1368	0.1406	0.1443	0.1480	0.1517
0.40	0.1554	0.1591	0.1628	0.1664	0.1700	0.1736	0.1772	0.1808	0.1844	0.1879
0.50	0.1915	0.1950	0.1985	0.2019	0.2054	0.2088	0.2123	0.2157	0.2190	0.2224
0.60	0.2257	0.2291	0.2324	0.2357	0.2389	0.2422	0.2454	0.2486	0.2517	0.2549
0.70	0.2580	0.2611	0.2642	0.2673	0.2704	0.2734	0.2764	0.2794	0.2823	0.2852
0.80	0.2881	0.2910	0.2939	0.2967	0.2995	0.3023	0.3051	0.3078	0.3106	0.3133
0.90	0.3159	0.3186	0.3212	0.3238	0.3264	0.3289	0.3315	0.3340	0.3365	0.3389
1.00	0.3413	0.3438	0.3461	0.3485	0.3508	0.3531	0.3554	0.3577	0.3599	0.3621
1.10	0.3643	0.3665	0.3686	0.3708	0.3729	0.3749	0.3770	0.3790	0.3810	0.3830
1.20	0.3849	0.3869	0.3888	0.3907	0.3925	0.3944	0.3962	0.3980	0.3997	0.4015
1.30	0.4032	0.4049	0.4066	0.4082	0.4099	0.4115	0.4131	0.4147	0.4162	0.4177
1.40	0.4192	0.4207	0.4222	0.4236	0.4251	0.4265	0.4279	0.4292	0.4306	0.4319
1.50	0.4332	0.4345	0.4357	0.4370	0.4382	0.4394	0.4406	0.4418	0.4429	0.4441
1.60	0.4452	0.4463	0.4474	0.4484	0.4495	0.4505	0.4515	0.4525	0.4535	0.4545
1.70	0.4554	0.4564	0.4573	0.4582	0.4591	0.4599	0.4608	0.4616	0.4625	0.4633
1.80	0.4641	0.4649	0.4656	0.4664	0.4671	0.4678	0.4686	0.4693	0.4699	0.4706
1.90	0.4713	0.4719	0.4726	0.4732	0.4738	0.4744	0.4750	0.4756	0.4761	0.4767
2.00	0.4772	0.4778	0.4783	0.4788	0.4793	0.4798	0.4803	0.4808	0.4812	0.4817
2.10	0.4821	0.4826	0.4830	0.4834	0.4838	0.4842	0.4846	0.4850	0.4854	0.4857
2.20	0.4861	0.4864	0.4868	0.4871	0.4875	0.4878	0.4881	0.4884	0.4887	0.4890
2.30	0.4893	0.4896	0.4898	0.4901	0.4904	0.4906	0.4909	0.4911	0.4913	0.4916
2.40	0.4918	0.4920	0.4922	0.4925	0.4927	0.4929	0.4931	0.4932	0.4934	0.4936
2.50	0.4938	0.4940	0.4941	0.4943	0.4945	0.4946	0.4948	0.4949	0.4951	0.4952

z										
2.60	0.4953	0.4955	0.4956	0.4957	0.4959	0.4960	0.4961	0.4962	0.4963	0.4964
2.70	0.4965	0.4966	0.4967	0.4968	0.4969	0.4970	0.4971	0.4972	0.4973	0.4974
2.80	0.4974	0.4975	0.4976	0.4977	0.4977	0.4978	0.4979	0.4979	0.4980	0.4981
2.90	0.4981	0.4982	0.4982	0.4983	0.4984	0.4984	0.4985	0.4985	0.4986	0.4986
3.00	0.4987	0.4987	0.4987	0.4988	0.4988	0.4989	0.4989	0.4989	0.4990	0.4990
3.10	0.4990	0.4991	0.4991	0.4991	0.4992	0.4992	0.4992	0.4992	0.4993	0.4993
3.20	0.4993	0.4993	0.4994	0.4994	0.4994	0.4994	0.4994	0.4995	0.4995	0.4995
3.30	0.4995	0.4995	0.4995	0.4996	0.4996	0.4996	0.4996	0.4996	0.4996	0.4997
3.40	0.4997	0.4997	0.4997	0.4997	0.4997	0.4997	0.4997	0.4997	0.4997	0.4998
3.50	0.4998	0.4998	0.4998	0.4998	0.4998	0.4998	0.4998	0.4998	0.4998	0.4998
3.60	0.4998	0.4998	0.4999	0.4999	0.4999	0.4999	0.4999	0.4999	0.4999	0.4999
3.70	0.4999	0.4999	0.4999	0.4999	0.4999	0.4999	0.4999	0.4999	0.4999	0.4999
3.80	0.4999	0.4999	0.4999	0.4999	0.4999	0.4999	0.4999	0.4999	0.4999	0.4999
3.90	0.5000	0.5000	0.5000	0.5000	0.5000	0.5000	0.5000	0.5000	0.5000	0.5000

Note: The table gives values of $P[0 < Z < z]$. If z is negative, then $P[Z < -z] = 0.50 - P[0 < Z < z]$.

TABLE A.2

t-Table

v^a	0.2	0.1	0.05	0.025	α 0.01	0.005	0.001	0.0005	0.0001
1	1.3764	3.0777	6.3138	12.706	31.821	63.657	318.31	636.62	3183.1
2	1.0607	1.8856	2.9200	4.3027	6.9646	9.9248	22.327	31.599	70.700
3	0.9785	1.6377	2.3534	3.1824	4.5407	5.8409	10.215	12.924	22.204
4	0.9410	1.5332	2.1318	2.7764	3.7470	4.6041	7.1732	8.6103	13.034
5	0.9195	1.4759	2.0150	2.5706	3.3649	4.0322	5.8934	6.8688	9.6776
6	0.9057	1.4398	1.9432	2.4469	3.1427	3.7074	5.2076	5.9588	8.0248
7	0.8960	1.4149	1.8946	2.3646	2.9980	3.4995	4.7853	5.4079	7.0634
8	0.8889	1.3968	1.8595	2.3060	2.8965	3.3554	4.5008	5.0413	6.4420
9	0.8834	1.3830	1.8331	2.2622	2.8214	3.2498	4.2968	4.7809	6.0101
10	0.8791	1.3722	1.8125	2.2281	2.7638	3.1693	4.1437	4.5869	5.6938
11	0.8755	1.3634	1.7959	2.2010	2.7181	3.1058	4.0247	4.4370	5.4528
12	0.8726	1.3562	1.7823	2.1788	2.6810	3.0545	3.9296	4.3178	5.2633
13	0.8702	1.3502	1.7709	2.1604	2.6503	3.0123	3.8520	4.2208	5.1106
14	0.8681	1.3450	1.7613	2.1448	2.6245	2.9768	3.7874	4.1405	4.9850
15	0.8662	1.3406	1.7531	2.1314	2.6025	2.9467	3.7328	4.0728	4.8800
16	0.8647	1.3368	1.7459	2.1199	2.5835	2.9208	3.6862	4.0150	4.7909
17	0.8633	1.3334	1.7396	2.1098	2.5669	2.8982	3.6458	3.9651	4.7144
18	0.8620	1.3304	1.7341	2.1009	2.5524	2.8784	3.6105	3.9216	4.6480
19	0.8610	1.3277	1.7291	2.0930	2.5395	2.8609	3.5794	3.8834	4.5899
20	0.8600	1.3253	1.7247	2.0860	2.5280	2.8453	3.5518	3.8495	4.5385
21	0.8591	1.3232	1.7207	2.0796	2.5176	2.8314	3.5272	3.8193	4.4929
22	0.8583	1.3212	1.7171	2.0739	2.5083	2.8188	3.5050	3.7921	4.4520
23	0.8575	1.3195	1.7139	2.0687	2.4999	2.8073	3.4850	3.7676	4.4152
24	0.8569	1.3178	1.7109	2.0639	2.4922	2.7969	3.4668	3.7454	4.3819
25	0.8562	1.3163	1.7081	2.0595	2.4851	2.7874	3.4502	3.7251	4.3517
26	0.8557	1.3150	1.7056	2.0555	2.4786	2.7787	3.4350	3.7066	4.3240
27	0.8551	1.3137	1.7033	2.0518	2.4727	2.7707	3.4210	3.6896	4.2987
28	0.8546	1.3125	1.7011	2.0484	2.4671	2.7633	3.4082	3.6739	4.2754
29	0.8542	1.3114	1.6991	2.0452	2.4620	2.7564	3.3962	3.6594	4.2539
30	0.8538	1.3104	1.6973	2.0423	2.4573	2.7500	3.3852	3.6460	4.2340
35	0.8520	1.3062	1.6896	2.0301	2.4377	2.7238	3.3400	3.5911	4.1531
40	0.8507	1.3031	1.6839	2.0211	2.4233	2.7045	3.3069	3.5510	4.0942
50	0.8489	1.2987	1.6759	2.0086	2.4033	2.6778	3.2614	3.4960	4.0140
60	0.8477	1.2958	1.6706	2.0003	2.3901	2.6603	3.2317	3.4602	3.9621
70	0.8468	1.2938	1.6669	1.9944	2.3808	2.6479	3.2108	3.4350	3.9257
80	0.8461	1.2922	1.6641	1.9901	2.3739	2.6387	3.1953	3.4163	3.8988
90	0.8456	1.2910	1.6620	1.9867	2.3685	2.6316	3.1833	3.4019	3.8780
100	0.8452	1.2901	1.6602	1.9840	2.3642	2.6259	3.1737	3.3905	3.8616
120	0.8446	1.2886	1.6577	1.9799	2.3578	2.6174	3.1595	3.3735	3.8372
∞	0.8416	1.2816	1.6449	1.9600	2.3263	2.5758	3.0902	3.2905	3.7190

[a] v = Degrees of freedom.

TABLE A.3a

F Table for $\alpha = 0.01$

v_2	v_1 = 1	2	3	4	5	6	7	8	9	10	12	15	20	25	30	40	60	120	∞
1	4052	4999	5403	5625	5764	5859	5928	5981	6023	6056	6106	6157	6209	6240	6261	6287	6313	6339	6366
2	98.50	99.00	99.17	99.25	99.30	99.33	99.36	99.37	99.39	99.40	99.42	99.43	99.45	99.46	99.47	99.47	99.48	99.49	99.50
3	34.12	30.82	29.46	28.71	28.24	27.91	27.67	27.49	27.35	27.23	27.05	26.87	26.69	26.58	26.50	26.41	26.32	26.22	26.13
4	21.20	18.00	16.69	15.98	15.52	15.21	14.98	14.80	14.66	14.55	14.37	14.20	14.02	13.91	13.84	13.75	13.65	13.56	13.46
5	16.26	13.27	12.06	11.39	10.97	10.67	10.46	10.29	10.16	10.05	9.89	9.72	9.55	9.45	9.38	9.29	9.20	9.11	9.02
6	13.75	10.92	9.78	9.15	8.75	8.47	8.26	8.10	7.98	7.87	7.72	7.56	7.40	7.30	7.23	7.14	7.06	6.97	6.88
7	12.25	9.55	8.45	7.85	7.46	7.19	6.99	6.84	6.72	6.62	6.47	6.31	6.16	6.06	5.99	5.91	5.82	5.74	5.65
8	11.26	8.65	7.59	7.01	6.63	6.37	6.18	6.03	5.91	5.81	5.67	5.52	5.36	5.26	5.20	5.12	5.03	4.95	4.86
9	10.56	8.02	6.99	6.42	6.06	5.80	5.61	5.47	5.35	5.26	5.11	4.96	4.81	4.71	4.65	4.57	4.48	4.40	4.31
10	10.04	7.56	6.55	5.99	5.64	5.39	5.20	5.06	4.94	4.85	4.71	4.56	4.41	4.31	4.25	4.17	4.08	4.00	3.91
11	9.65	7.21	6.22	5.67	5.32	5.07	4.89	4.74	4.63	4.54	4.40	4.25	4.10	4.01	3.94	3.86	3.78	3.69	3.60
12	9.33	6.93	5.95	5.41	5.06	4.82	4.64	4.50	4.39	4.30	4.16	4.01	3.86	3.76	3.70	3.62	3.54	3.45	3.36
13	9.07	6.70	5.74	5.21	4.86	4.62	4.44	4.30	4.19	4.10	3.96	3.82	3.66	3.57	3.51	3.43	3.34	3.25	3.17
14	8.86	6.51	5.56	5.04	4.69	4.46	4.28	4.14	4.03	3.94	3.80	3.66	3.51	3.41	3.35	3.27	3.18	3.09	3.00
15	8.68	6.36	5.42	4.89	4.56	4.32	4.14	4.00	3.89	3.80	3.67	3.52	3.37	3.28	3.21	3.13	3.05	2.96	2.87
16	8.53	6.23	5.29	4.77	4.44	4.20	4.03	3.89	3.78	3.69	3.55	3.41	3.26	3.16	3.10	3.02	2.93	2.84	2.75
17	8.40	6.11	5.18	4.67	4.34	4.10	3.93	3.79	3.68	3.59	3.46	3.31	3.16	3.07	3.00	2.92	2.83	2.75	2.65
18	8.29	6.01	5.09	4.58	4.25	4.01	3.84	3.71	3.60	3.51	3.37	3.23	3.08	2.98	2.92	2.84	2.75	2.66	2.57
19	8.18	5.93	5.01	4.50	4.17	3.94	3.77	3.63	3.52	3.43	3.30	3.15	3.00	2.91	2.84	2.76	2.67	2.58	2.49
20	8.10	5.85	4.94	4.43	4.10	3.87	3.70	3.56	3.46	3.37	3.23	3.09	2.94	2.84	2.78	2.69	2.61	2.52	2.42
21	8.02	5.78	4.87	4.37	4.04	3.81	3.64	3.51	3.40	3.31	3.17	3.03	2.88	2.79	2.72	2.64	2.55	2.46	2.36
22	7.95	5.72	4.82	4.31	3.99	3.76	3.59	3.45	3.35	3.26	3.12	2.98	2.83	2.73	2.67	2.58	2.50	2.40	2.31
23	7.88	5.66	4.76	4.26	3.94	3.71	3.54	3.41	3.30	3.21	3.07	2.93	2.78	2.69	2.62	2.54	2.45	2.35	2.26
24	7.82	5.61	4.72	4.22	3.90	3.67	3.50	3.36	3.26	3.17	3.03	2.89	2.74	2.64	2.58	2.49	2.40	2.31	2.21
25	7.77	5.57	4.68	4.18	3.85	3.63	3.46	3.32	3.22	3.13	2.99	2.85	2.70	2.60	2.54	2.45	2.36	2.27	2.17
26	7.72	5.53	4.64	4.14	3.82	3.59	3.42	3.29	3.18	3.09	2.96	2.81	2.66	2.57	2.50	2.42	2.33	2.23	2.13
27	7.68	5.49	4.60	4.11	3.78	3.56	3.39	3.26	3.15	3.06	2.93	2.78	2.63	2.54	2.47	2.38	2.29	2.20	2.10

TABLE A.3a

F Table for $\alpha = 0.01$ (continued)

v_2^b	v_1^a 1	2	3	4	5	6	7	8	9	10	12	15	20	25	30	40	60	120	∞
28	7.64	5.45	4.57	4.07	3.75	3.53	3.36	3.23	3.12	3.03	2.90	2.75	2.60	2.51	2.44	2.35	2.26	2.17	2.06
29	7.60	5.42	4.54	4.04	3.73	3.50	3.33	3.20	3.09	3.00	2.87	2.73	2.57	2.48	2.41	2.33	2.23	2.14	2.03
30	7.56	5.39	4.51	4.02	3.70	3.47	3.30	3.17	3.07	2.98	2.84	2.70	2.55	2.45	2.39	2.30	2.21	2.11	2.01
35	7.42	5.27	4.40	3.91	3.59	3.37	3.20	3.07	2.96	2.88	2.74	2.60	2.44	2.35	2.28	2.19	2.10	2.00	1.89
40	7.31	5.18	4.31	3.83	3.51	3.29	3.12	2.99	2.89	2.80	2.66	2.52	2.37	2.27	2.20	2.11	2.02	1.92	1.80
45	7.23	5.11	4.25	3.77	3.45	3.23	3.07	2.94	2.83	2.74	2.61	2.46	2.31	2.21	2.14	2.05	1.96	1.85	1.74
50	7.17	5.06	4.20	3.72	3.41	3.19	3.02	2.89	2.78	2.70	2.56	2.42	2.27	2.17	2.10	2.01	1.91	1.80	1.68
60	7.08	4.98	4.13	3.65	3.34	3.12	2.95	2.82	2.72	2.63	2.50	2.35	2.20	2.10	2.03	1.94	1.84	1.73	1.60
70	7.01	4.92	4.07	3.60	3.29	3.07	2.91	2.78	2.67	2.59	2.45	2.31	2.15	2.05	1.98	1.89	1.78	1.67	1.54
80	6.96	4.88	4.04	3.56	3.26	3.04	2.87	2.74	2.64	2.55	2.42	2.27	2.12	2.01	1.94	1.85	1.75	1.63	1.49
90	6.93	4.85	4.01	3.53	3.23	3.01	2.84	2.72	2.61	2.52	2.39	2.24	2.09	1.99	1.92	1.82	1.72	1.60	1.46
100	6.90	4.82	3.98	3.51	3.21	2.99	2.82	2.69	2.59	2.50	2.37	2.22	2.07	1.97	1.89	1.80	1.69	1.57	1.43
110	6.87	4.80	3.96	3.49	3.19	2.97	2.81	2.68	2.57	2.49	2.35	2.21	2.05	1.95	1.88	1.78	1.67	1.55	1.40
120	6.85	4.79	3.95	3.48	3.17	2.96	2.79	2.66	2.56	2.47	2.34	2.19	2.03	1.93	1.86	1.76	1.66	1.53	1.38
∞	6.63	4.61	3.78	3.32	3.02	2.80	2.64	2.51	2.41	2.32	2.18	2.04	1.88	1.77	1.70	1.59	1.47	1.32	1.00

[a] v_1 = Numerator degrees of freedom.
[b] v_2 = Denominator degrees of freedom.

TABLE A.3b

F Table for $\alpha = 0.05$

v_2^b	v_1^a																		
	1	2	3	4	5	6	7	8	9	10	12	15	20	25	30	40	60	120	∞
1	161.5	199.5	215.7	224.6	230.2	234.0	236.8	238.9	240.5	241.9	243.9	246.0	248.0	249.3	250.1	251.1	252.2	253.3	254.3
2	18.51	19.00	19.16	19.25	19.30	19.33	19.35	19.37	19.38	19.40	19.41	19.43	19.45	19.46	19.46	19.47	19.48	19.49	19.50
3	10.13	9.55	9.28	9.12	9.01	8.94	8.89	8.85	8.81	8.79	8.74	8.70	8.66	8.63	8.62	8.59	8.57	8.55	8.53
4	7.71	6.94	6.59	6.39	6.26	6.16	6.09	6.04	6.00	5.96	5.91	5.86	5.80	5.77	5.75	5.72	5.69	5.66	5.63
5	6.61	5.79	5.41	5.19	5.05	4.95	4.88	4.82	4.77	4.74	4.68	4.62	4.56	4.52	4.50	4.46	4.43	4.40	4.36
6	5.99	5.14	4.76	4.53	4.39	4.28	4.21	4.15	4.10	4.06	4.00	3.94	3.87	3.83	3.81	3.77	3.74	3.70	3.67
7	5.59	4.74	4.35	4.12	3.97	3.87	3.79	3.73	3.68	3.64	3.57	3.51	3.44	3.40	3.38	3.34	3.30	3.27	3.23
8	5.32	4.46	4.07	3.84	3.69	3.58	3.50	3.44	3.39	3.35	3.28	3.22	3.15	3.11	3.08	3.04	3.01	2.97	2.93
9	5.12	4.26	3.86	3.63	3.48	3.37	3.29	3.23	3.18	3.14	3.07	3.01	2.94	2.89	2.86	2.83	2.79	2.75	2.71
10	4.96	4.10	3.71	3.48	3.33	3.22	3.14	3.07	3.02	2.98	2.91	2.85	2.77	2.73	2.70	2.66	2.62	2.58	2.54
11	4.84	3.98	3.59	3.36	3.20	3.09	3.01	2.95	2.90	2.85	2.79	2.72	2.65	2.60	2.57	2.53	2.49	2.45	2.40
12	4.75	3.89	3.49	3.26	3.11	3.00	2.91	2.85	2.80	2.75	2.69	2.62	2.54	2.50	2.47	2.43	2.38	2.34	2.30
13	4.67	3.81	3.41	3.18	3.03	2.92	2.83	2.77	2.71	2.67	2.60	2.53	2.46	2.41	2.38	2.34	2.30	2.25	2.21
14	4.60	3.74	3.34	3.11	2.96	2.85	2.76	2.70	2.65	2.60	2.53	2.46	2.39	2.34	2.31	2.27	2.22	2.18	2.13
15	4.54	3.68	3.29	3.06	2.90	2.79	2.71	2.64	2.59	2.54	2.48	2.40	2.33	2.28	2.25	2.20	2.16	2.11	2.07
16	4.49	3.63	3.24	3.01	2.85	2.74	2.66	2.59	2.54	2.49	2.42	2.35	2.28	2.23	2.19	2.15	2.11	2.06	2.01
17	4.45	3.59	3.20	2.96	2.81	2.70	2.61	2.55	2.49	2.45	2.38	2.31	2.23	2.18	2.15	2.10	2.06	2.01	1.96
18	4.41	3.55	3.16	2.93	2.77	2.66	2.58	2.51	2.46	2.41	2.34	2.27	2.19	2.14	2.11	2.06	2.02	1.97	1.92
19	4.38	3.52	3.13	2.90	2.74	2.63	2.54	2.48	2.42	2.38	2.31	2.23	2.16	2.11	2.07	2.03	1.98	1.93	1.88
20	4.35	3.49	3.10	2.87	2.71	2.60	2.51	2.45	2.39	2.35	2.28	2.20	2.12	2.07	2.04	1.99	1.95	1.90	1.84
21	4.32	3.47	3.07	2.84	2.68	2.57	2.49	2.42	2.37	2.32	2.25	2.18	2.10	2.05	2.01	1.96	1.92	1.87	1.81
22	4.30	3.44	3.05	2.82	2.66	2.55	2.46	2.40	2.34	2.30	2.23	2.15	2.07	2.02	1.98	1.94	1.89	1.84	1.78
23	4.28	3.42	3.03	2.80	2.64	2.53	2.44	2.37	2.32	2.27	2.20	2.13	2.05	2.00	1.96	1.91	1.86	1.81	1.76
24	4.26	3.40	3.01	2.78	2.62	2.51	2.42	2.36	2.30	2.25	2.18	2.11	2.03	1.97	1.94	1.89	1.84	1.79	1.73
25	4.24	3.39	2.99	2.76	2.60	2.49	2.40	2.34	2.28	2.24	2.16	2.09	2.01	1.96	1.92	1.87	1.82	1.77	1.71
26	4.23	3.37	2.98	2.74	2.59	2.47	2.39	2.32	2.27	2.22	2.15	2.07	1.99	1.94	1.90	1.85	1.80	1.75	1.69
27	4.21	3.35	2.96	2.73	2.57	2.46	2.37	2.31	2.25	2.20	2.13	2.06	1.97	1.92	1.88	1.84	1.79	1.73	1.67

TABLE A.3b

F Table for $\alpha = 0.05$ (continued)

ν_2	ν_1 1	2	3	4	5	6	7	8	9	10	12	15	20	25	30	40	60	120	∞
28	4.20	3.34	2.95	2.71	2.56	2.45	2.36	2.29	2.24	2.19	2.12	2.04	1.96	1.91	1.87	1.82	1.77	1.71	1.65
29	4.18	3.33	2.93	2.70	2.55	2.43	2.35	2.28	2.22	2.18	2.10	2.03	1.94	1.89	1.85	1.81	1.75	1.70	1.64
30	4.17	3.32	2.92	2.69	2.53	2.42	2.33	2.27	2.21	2.16	2.09	2.01	1.93	1.88	1.84	1.79	1.74	1.68	1.62
35	4.12	3.27	2.87	2.64	2.49	2.37	2.29	2.22	2.16	2.11	2.04	1.96	1.88	1.82	1.79	1.74	1.68	1.62	1.56
40	4.08	3.23	2.84	2.61	2.45	2.34	2.25	2.18	2.12	2.08	2.00	1.92	1.84	1.78	1.74	1.69	1.64	1.58	1.51
45	4.06	3.20	2.81	2.58	2.42	2.31	2.22	2.15	2.10	2.05	1.97	1.89	1.81	1.75	1.71	1.66	1.60	1.54	1.47
50	4.03	3.18	2.79	2.56	2.40	2.29	2.20	2.13	2.07	2.03	1.95	1.87	1.78	1.73	1.69	1.63	1.58	1.51	1.44
60	4.00	3.15	2.76	2.53	2.37	2.25	2.17	2.10	2.04	1.99	1.92	1.84	1.75	1.69	1.65	1.59	1.53	1.47	1.39
70	3.98	3.13	2.74	2.50	2.35	2.23	2.14	2.07	2.02	1.97	1.89	1.81	1.72	1.66	1.62	1.57	1.50	1.44	1.35
80	3.96	3.11	2.72	2.49	2.33	2.21	2.13	2.06	2.00	1.95	1.88	1.79	1.70	1.64	1.60	1.54	1.48	1.41	1.32
90	3.95	3.10	2.71	2.47	2.32	2.20	2.11	2.04	1.99	1.94	1.86	1.78	1.69	1.63	1.59	1.53	1.46	1.39	1.30
100	3.94	3.09	2.70	2.46	2.31	2.19	2.10	2.03	1.97	1.93	1.85	1.77	1.68	1.62	1.57	1.52	1.45	1.38	1.28
110	3.93	3.08	2.69	2.45	2.30	2.18	2.09	2.02	1.97	1.92	1.84	1.76	1.67	1.61	1.56	1.50	1.44	1.36	1.27
120	3.92	3.07	2.68	2.45	2.29	2.18	2.09	2.02	1.96	1.91	1.83	1.75	1.66	1.60	1.55	1.50	1.43	1.35	1.25
∞	3.84	3.00	2.60	2.37	2.21	2.10	2.01	1.94	1.88	1.83	1.75	1.67	1.57	1.51	1.46	1.39	1.32	1.22	1.00

[a] ν_1 = Numerator degrees of freedom.

[b] ν_2 = Denominator degrees of freedom.

TABLE A.4

Constants Used for Estimation and Construction of Control Charts

n	c_4	d_2	d_3	A	A_2	A_3	B_3	B_4	B_5	B_6	D_1	D_2	D_3	D_4
2	0.7979	1.1280	0.8525	2.1213	1.8806	2.6586	0	3.2664	0	2.6063	0	3.6855	0	3.2672
3	0.8862	1.6929	0.8884	1.7321	1.0231	1.9545	0	2.5684	0	2.2761	0	4.3581	0	2.5743
4	0.9213	2.0589	0.8798	1.5000	0.7286	1.6281	0	2.2662	0	2.0879	0	4.6983	0	2.2819
5	0.9399	2.3261	0.8641	1.3416	0.5768	1.4273	0	2.0889	0	1.9635	0	4.9184	0	2.1144
6	0.9516	2.5342	0.8480	1.2247	0.4833	1.2872	0.0302	1.9698	0.0286	1.8744	0	5.0782	0	2.0039
7	0.9593	2.7042	0.8332	1.1339	0.4193	1.1819	0.1182	1.8818	0.1133	1.8055	0.2046	5.2038	0.0756	1.9244
8	0.9651	2.8474	0.8198	1.0607	0.3725	1.0991	0.1847	1.8153	0.1783	1.7517	0.3880	5.3068	0.1363	1.8637
9	0.9693	2.9700	0.8078	1.0000	0.3367	1.0317	0.2389	1.7611	0.2317	1.7069	0.5466	5.3934	0.1840	1.8160
10	0.9727	3.0779	0.7971	0.9487	0.3082	0.9753	0.2843	1.7157	0.2765	1.6689	0.6866	5.4692	0.2231	1.7769
11	0.9753	3.1726	0.7873	0.9045	0.2851	0.9273	0.3221	1.6779	0.3141	1.6367	0.8107	5.5345	0.2555	1.7445
12	0.9776	3.2584	0.7785	0.8660	0.2658	0.8859	0.3541	1.6459	0.3462	1.6090	0.9229	5.5939	0.2832	1.7168
13	0.9794	3.3356	0.7704	0.8321	0.2494	0.8496	0.3815	1.6185	0.3736	1.5852	1.0244	5.6468	0.3071	1.6929
14	0.9810	3.4072	0.7630	0.8018	0.2353	0.8173	0.4067	1.5933	0.3990	1.5630	1.1182	5.6962	0.3282	1.6718
15	0.9823	3.4722	0.7562	0.7746	0.2231	0.7886	0.4279	1.5721	0.4204	1.5442	1.2036	5.7408	0.3466	1.6534
16	0.9835	3.5323	0.7499	0.7500	0.2123	0.7626	0.4482	1.5518	0.4408	1.5262	1.2826	5.7820	0.3631	1.6369
17	0.9845	3.5881	0.7441	0.7276	0.2028	0.7391	0.4656	1.5344	0.4583	1.5107	1.3558	5.8204	0.3779	1.6221
18	0.9854	3.6403	0.7386	0.7071	0.1942	0.7176	0.4817	1.5183	0.4746	1.4962	1.4245	5.8561	0.3913	1.6087
19	0.9862	3.6887	0.7335	0.6882	0.1866	0.6979	0.4964	1.5036	0.4895	1.4829	1.4882	5.8892	0.4034	1.5966
20	0.9870	3.7355	0.7287	0.6708	0.1796	0.6797	0.5096	1.4904	0.5029	1.4709	1.5494	5.9216	0.4148	1.5852
21	0.9875	3.7779	0.7242	0.6547	0.1733	0.6629	0.5231	1.4769	0.5166	1.4586	1.6053	5.9505	0.4249	1.5751
22	0.9882	3.8197	0.7199	0.6396	0.1674	0.6472	0.5349	1.4651	0.5287	1.4477	1.6600	5.9794	0.4346	1.5654
23	0.9887	3.8580	0.7159	0.6255	0.1621	0.6327	0.5451	1.4549	0.5390	1.4384	1.7103	6.0057	0.4433	1.5567
24	0.9892	3.8956	0.7121	0.6124	0.1572	0.6191	0.5555	1.4445	0.5495	1.4289	1.7593	6.0319	0.4516	1.5484
25	0.9896	3.9308	0.7084	0.6000	0.1526	0.6063	0.5639	1.4361	0.5581	1.4211	1.8056	6.0560	0.4593	1.5407

TABLE A.5

Extended Standard Normal Tables

z Value	Cum. Prob	PPM	z Value	Cum. Prob	PPM
−3.00	0.0013499672	1349.9672	−3.56	0.0001854674	185.4674
−3.01	0.0013063077	1306.3077	−3.57	0.0001785299	178.5299
−3.02	0.0012639426	1263.9426	−3.58	0.0001718356	171.8356
−3.03	0.0012228379	1222.8379	−3.59	0.0001653768	165.3768
−3.04	0.0011829598	1182.9598	−3.60	0.0001591457	159.1457
−3.05	0.0011442758	1144.2758	−3.61	0.0001531349	153.1349
−3.06	0.0011067538	1106.7538	−3.62	0.0001473372	147.3372
−3.07	0.0010703626	1070.3626	−3.63	0.0001417457	141.7457
−3.08	0.0010350715	1035.0715	−3.64	0.0001363534	136.3534
−3.09	0.0010008508	1000.8508	−3.65	0.0001311538	131.1538
−3.10	0.0009676712	967.6712	−3.66	0.0001261406	126.1406
−3.11	0.0009355045	935.5045	−3.67	0.0001213076	121.3076
−3.12	0.0009043226	904.3226	−3.68	0.0001166487	116.6487
−3.13	0.0008740986	874.0986	−3.69	0.0001121581	112.1581
−3.14	0.0008448059	844.8059	−3.70	0.0001078301	107.8301
−3.15	0.0008164187	816.4187	−3.71	0.0001036594	103.6594
−3.16	0.0007889117	788.9117	−3.72	0.0000996405	99.6405
−3.17	0.0007622602	762.2602	−3.73	0.0000957684	95.7684
−3.18	0.0007364404	736.4404	−3.74	0.0000920380	92.0380
−3.19	0.0007114286	711.4286	−3.75	0.0000884446	88.4446
−3.20	0.0006872021	687.2021	−3.76	0.0000849834	84.9834
−3.21	0.0006637385	663.7385	−3.77	0.0000816499	81.6499
−3.22	0.0006410161	641.0161	−3.78	0.0000784397	78.4397
−3.23	0.0006190137	619.0137	−3.79	0.0000753486	75.3486
−3.24	0.0005977105	597.7105	−3.80	0.0000723724	72.3724
−3.25	0.0005770865	577.0865	−3.81	0.0000695072	69.5072
−3.26	0.0005571219	557.1219	−3.82	0.0000667491	66.7491
−3.27	0.0005377977	537.7977	−3.83	0.0000640944	64.0944
−3.28	0.0005190951	519.0951	−3.84	0.0000615394	61.5394
−3.29	0.0005009959	500.9959	−3.85	0.0000590806	59.0806
−3.30	0.0004834825	483.4825	−3.86	0.0000567147	56.7147
−3.31	0.0004665376	466.5376	−3.87	0.0000544383	54.4383
−3.32	0.0004501443	450.1443	−3.88	0.0000522484	52.2484
−3.33	0.0004342863	434.2863	−3.89	0.0000501418	50.1418
−3.34	0.0004189477	418.9477	−3.90	0.0000481155	48.1155
−3.35	0.0004041129	404.1129	−3.91	0.0000461668	46.1668
−3.36	0.0003897667	389.7667	−3.92	0.0000442927	44.2927
−3.37	0.0003758946	375.8946	−3.93	0.0000424907	42.4907
−3.38	0.0003624821	362.4821	−3.94	0.0000407581	40.7581
−3.39	0.0003495154	349.5154	−3.95	0.0000390925	39.0925
−3.40	0.0003369808	336.9808	−3.96	0.0000374913	37.4913
−3.41	0.0003248652	324.8652	−3.97	0.0000359523	35.9523
−3.42	0.0003131558	313.1558	−3.98	0.0000344732	34.4732
−3.43	0.0003018400	301.8400	−3.99	0.0000330518	33.0518
−3.44	0.0002909058	290.9058	−4.00	0.0000316860	31.6860

TABLE A.5

Extended Standard Normal Tables (continued)

z Value	Cum. Prob	PPM	z Value	Cum. Prob	PPM
−3.45	0.0002803412	280.3412	−4.01	0.0000303738	30.3738
−3.46	0.0002701349	270.1349	−4.02	0.0000291131	29.1131
−3.47	0.0002602757	260.2757	−4.03	0.0000279021	27.9021
−3.48	0.0002507526	250.7526	−4.04	0.0000267389	26.7389
−3.49	0.0002415553	241.5553	−4.05	0.0000256217	25.6217
−3.50	0.0002326734	232.6734	−4.06	0.0000245489	24.5489
−3.51	0.0002240969	224.0969	−4.07	0.0000235188	23.5188
−3.52	0.0002158162	215.8162	−4.08	0.0000225297	22.5297
−3.53	0.0002078219	207.8219	−4.09	0.0000215802	21.5802
−3.54	0.0002001049	200.1049	−4.10	0.0000206687	20.6687
−3.55	0.0001926562	192.6562	−4.11	0.0000197938	19.7938
−4.12	0.0000189542	18.9542	−4.69	0.0000013676	1.3676
−4.13	0.0000181484	18.1484	−4.70	0.0000013023	1.3023
−4.14	0.0000173753	17.3753	−4.71	0.0000012400	1.2400
−4.15	0.0000166335	16.6335	−4.72	0.0000011806	1.1806
−4.16	0.0000159218	15.9218	−4.73	0.0000011239	1.1239
−4.17	0.0000152391	15.2391	−4.74	0.0000010699	1.0699
−4.18	0.0000145843	14.5843	−4.75	0.0000010183	1.0183
−4.19	0.0000139563	13.9563	−4.76	0.0000009692	0.9692
−4.20	0.0000133541	13.3541	−4.77	0.0000009223	0.9223
−4.21	0.0000127766	12.7766	−4.78	0.0000008776	0.8776
−4.22	0.0000122230	12.2230	−4.79	0.0000008350	0.8350
−4.23	0.0000116922	11.6922	−4.80	0.0000007944	0.7944
−4.24	0.0000111834	11.1834	−4.81	0.0000007556	0.7556
−4.25	0.0000106957	10.6957	−4.82	0.0000007187	0.7187
−4.26	0.0000102283	10.2283	−4.83	0.0000006836	0.6836
−4.27	0.0000097804	9.7804	−4.84	0.0000006501	0.6501
−4.28	0.0000093512	9.3512	−4.85	0.0000006181	0.6181
−4.29	0.0000089400	8.9400	−4.86	0.0000005877	0.5877
−4.30	0.0000085460	8.5460	−4.87	0.0000005588	0.5588
−4.31	0.0000081687	8.1687	−4.88	0.0000005312	0.5312
−4.32	0.0000078072	7.8072	−4.89	0.0000005049	0.5049
−4.33	0.0000074610	7.4610	−4.90	0.0000004799	0.4799
−4.34	0.0000071295	7.1295	−4.91	0.0000004560	0.4560
−4.35	0.0000068121	6.8121	−4.92	0.0000004334	0.4334
−4.36	0.0000065082	6.5082	−4.93	0.0000004118	0.4118
−4.37	0.0000062172	6.2172	−4.94	0.0000003912	0.3912
−4.38	0.0000059387	5.9387	−4.95	0.0000003716	0.3716
−4.39	0.0000056721	5.6721	−4.96	0.0000003530	0.3530
−4.40	0.0000054170	5.4170	−4.97	0.0000003353	0.3353
−4.41	0.0000051728	5.1728	−4.98	0.0000003184	0.3184
−4.42	0.0000049392	4.9392	−4.99	0.0000003024	0.3024
−4.43	0.0000047156	4.7156	−5.00	0.0000002871	0.2871
−4.44	0.0000045018	4.5018	−5.01	0.0000002726	0.2726

(Continued)

TABLE A.5

Extended Standard Normal Tables (continued)

z Value	Cum. Prob	PPM	z Value	Cum. Prob	PPM
−4.45	0.0000042972	4.2972	−5.02	0.0000002588	0.2588
−4.46	0.0000041016	4.1016	−5.03	0.0000002456	0.2456
−4.47	0.0000039145	3.9145	−5.04	0.0000002331	0.2331
−4.48	0.0000037355	3.7355	−5.05	0.0000002213	0.2213
−4.49	0.0000035644	3.5644	−5.06	0.0000002100	0.2100
−4.50	0.0000034008	3.4008	−5.07	0.0000001992	0.1992
−4.51	0.0000032444	3.2444	−5.08	0.0000001890	0.1890
−4.52	0.0000030949	3.0949	−5.09	0.0000001793	0.1793
−4.53	0.0000029520	2.9520	−5.10	0.0000001701	0.1701
−4.54	0.0000028154	2.8154	−5.11	0.0000001614	0.1614
−4.55	0.0000026849	2.6849	−5.12	0.0000001530	0.1530
−4.56	0.0000025602	2.5602	−5.13	0.0000001451	0.1451
−4.57	0.0000024411	2.4411	−5.14	0.0000001376	0.1376
−4.58	0.0000023272	2.3272	−5.15	0.0000001305	0.1305
−4.59	0.0000022185	2.2185	−5.11	0.0000001614	0.1614
−4.60	0.0000021146	2.1146	−5.12	0.0000001530	0.1530
−4.61	0.0000020155	2.0155	−5.13	0.0000001451	0.1451
−4.62	0.0000019207	1.9207	−5.14	0.0000001376	0.1376
−4.63	0.0000018303	1.8303	−5.15	0.0000001305	0.1305
−4.64	0.0000017439	1.7439	−5.16	0.0000001237	0.1237
−4.65	0.0000016615	1.6615	−5.17	0.0000001173	0.1173
−4.66	0.0000015828	1.5828	−5.18	0.0000001112	0.1112
−4.67	0.0000015077	1.5077	−5.19	0.0000001053	0.1053
−4.68	0.0000014360	1.4360	−5.20	0.0000000998	0.0998
−5.21	0.0000000946	0.0946	−5.61	0.0000000101	0.0101
−5.22	0.0000000896	0.0896	−5.62	0.0000000096	0.0096
−5.23	0.0000000849	0.0849	−5.63	0.0000000090	0.0090
−5.24	0.0000000804	0.0804	−5.64	0.0000000085	0.0085
−5.25	0.0000000762	0.0762	−5.65	0.0000000080	0.0080
−5.26	0.0000000722	0.0722	−5.66	0.0000000076	0.0076
−5.27	0.0000000684	0.0684	−5.67	0.0000000072	0.0072
−5.28	0.0000000647	0.0647	−5.68	0.0000000068	0.0068
−5.29	0.0000000613	0.0613	−5.69	0.0000000064	0.0064
−5.30	0.0000000580	0.0580	−5.70	0.0000000060	0.0060
−5.31	0.0000000549	0.0549	−5.71	0.0000000057	0.0057
−5.32	0.0000000520	0.0520	−5.72	0.0000000053	0.0053
−5.33	0.0000000492	0.0492	−5.73	0.0000000050	0.0050
−5.34	0.0000000466	0.0466	−5.74	0.0000000047	0.0047
−5.35	0.0000000441	0.0441	−5.75	0.0000000045	0.0045
−5.36	0.0000000417	0.0417	−5.76	0.0000000042	0.0042
−5.37	0.0000000395	0.0395	−5.77	0.0000000040	0.0040
−5.38	0.0000000373	0.0373	−5.78	0.0000000037	0.0037
−5.39	0.0000000353	0.0353	−5.79	0.0000000035	0.0035
−5.40	0.0000000334	0.0334	−5.80	0.0000000033	0.0033
−5.41	0.0000000316	0.0316	−5.81	0.0000000031	0.0031

TABLE A.5

Extended Standard Normal Tables (continued)

z Value	Cum. Prob	PPM	z Value	Cum. Prob	PPM
−5.42	0.0000000299	0.0299	−5.82	0.0000000030	0.0030
−5.43	0.0000000282	0.0282	−5.83	0.0000000028	0.0028
−5.44	0.0000000267	0.0267	−5.84	0.0000000026	0.0026
−5.45	0.0000000252	0.0252	−5.85	0.0000000025	0.0025
−5.46	0.0000000239	0.0239	−5.86	0.0000000023	0.0023
−5.47	0.0000000226	0.0226	−5.87	0.0000000022	0.0022
−5.48	0.0000000213	0.0213	−5.88	0.0000000021	0.0021
−5.49	0.0000000201	0.0201	−5.89	0.0000000019	0.0019
−5.50	0.0000000190	0.0190	−5.90	0.0000000018	0.0018
−5.51	0.0000000180	0.0180	−5.91	0.0000000017	0.0017
−5.52	0.0000000170	0.0170	−5.92	0.0000000016	0.0016
−5.53	0.0000000161	0.0161	−5.93	0.0000000015	0.0015
−5.54	0.0000000152	0.0152	−5.94	0.0000000014	0.0014
−5.55	0.0000000143	0.0143	−5.95	0.0000000013	0.0013
−5.56	0.0000000135	0.0135	−5.96	0.0000000013	0.0013
−5.57	0.0000000128	0.0128	−5.97	0.0000000012	0.0012
−5.58	0.0000000121	0.0121	−5.98	0.0000000011	0.0011
−5.59	0.0000000114	0.0114	−5.99	0.0000000011	0.0011
−5.60	0.0000000107	0.0107	−6.00	0.0000000010	0.0010

TABLE A.6
Chi-Square Table

v	α (Area to the Right)											
	0.999	0.995	0.99	0.975	0.95	0.9	0.1	0.05	0.025	0.01	0.005	0.001
1	0.0000	0.0000	0.0002	0.0010	0.0039	0.0158	2.7055	3.8415	5.0239	6.6349	7.8794	10.8276
2	0.0020	0.0100	0.0201	0.0506	0.1026	0.2107	4.6052	5.9915	7.3778	9.2103	10.5966	13.8155
3	0.0243	0.0717	0.1148	0.2158	0.3518	0.5844	6.2514	7.8147	9.3484	11.3449	12.8382	16.2662
4	0.0908	0.2070	0.2971	0.4844	0.7107	1.0636	7.7794	9.4877	11.1433	13.2767	14.8603	18.4668
5	0.2102	0.4117	0.5543	0.8312	1.1455	1.6103	9.2364	11.0705	12.8325	15.0863	16.7496	20.5150
6	0.3811	0.6757	0.8721	1.2373	1.6354	2.2041	10.6446	12.5916	14.4494	16.8119	18.5476	22.4577
7	0.5985	0.9893	1.2390	1.6899	2.1673	2.8331	12.0170	14.0671	16.0128	18.4753	20.2777	24.3219
8	0.8571	1.3444	1.6465	2.1797	2.7326	3.4895	13.3616	15.5073	17.5345	20.0902	21.9550	26.1245
9	1.1519	1.7349	2.0879	2.7004	3.3251	4.1682	14.6837	16.9190	19.0228	21.6660	23.5894	27.8772
10	1.4787	2.1559	2.5582	3.2470	3.9403	4.8652	15.9872	18.3070	20.4832	23.2093	25.1882	29.5883
11	1.8339	2.6032	3.0535	3.8157	4.5748	5.5778	17.2750	19.6751	21.9200	24.7250	26.7568	31.2641
12	2.2142	3.0738	3.5706	4.4038	5.2260	6.3038	18.5493	21.0261	23.3367	26.2170	28.2995	32.9095
13	2.6172	3.5650	4.1069	5.0088	5.8919	7.0415	19.8119	22.3620	24.7356	27.6882	29.8195	34.5282
14	3.0407	4.0747	4.6604	5.6287	6.5706	7.7895	21.0641	23.6848	26.1189	29.1412	31.3193	36.1233
15	3.4827	4.6009	5.2293	6.2621	7.2609	8.5468	22.3071	24.9958	27.4884	30.5779	32.8013	37.6973
16	3.9416	5.1422	5.8122	6.9077	7.9616	9.3122	23.5418	26.2962	28.8454	31.9999	34.2672	39.2524
17	4.4161	5.6972	6.4078	7.5642	8.6718	10.0852	24.7690	27.5871	30.1910	33.4087	35.7185	40.7902
18	4.9048	6.2648	7.0149	8.2307	9.3905	10.8649	25.9894	28.8693	31.5264	34.8053	37.1565	42.3124
19	5.4068	6.8440	7.6327	8.9065	10.1170	11.6509	27.2036	30.1435	32.8523	36.1909	38.5823	43.8202
20	5.9210	7.4338	8.2604	9.5908	10.8508	12.4426	28.4120	31.4104	34.1696	37.5662	39.9968	45.3147
21	6.4467	8.0337	8.8972	10.2829	11.5913	13.2396	29.6151	32.6706	35.4789	38.9322	41.4011	46.7970
22	6.9830	8.6427	9.5425	10.9823	12.3380	14.0415	30.8133	33.9244	36.7807	40.2894	42.7957	48.2679
23	7.5292	9.2604	10.1957	11.6886	13.0905	14.8480	32.0069	35.1725	38.0756	41.6384	44.1813	49.7282
24	8.0849	9.8862	10.8564	12.4012	13.8484	15.6587	33.1962	36.4150	39.3641	42.9798	45.5585	51.1786

v												
25	8.6493	10.5197	11.5240	13.1197	14.6114	16.4734	34.3816	37.6525	40.6465	44.3141	46.9279	52.6197
26	9.2221	11.1602	12.1981	13.8439	15.3792	17.2919	35.5632	38.8851	41.9232	45.6417	48.2899	54.0520
27	9.8028	11.8076	12.8785	14.5734	16.1514	18.1139	36.7412	40.1133	43.1945	46.9629	49.6449	55.4760
28	10.3909	12.4613	13.5647	15.3079	16.9279	18.9392	37.9159	41.3371	44.4608	48.2782	50.9934	56.8923
29	10.9861	13.1211	14.2565	16.0471	17.7084	19.7677	39.0875	42.5570	45.7223	49.5879	52.3356	58.3012
30	11.5880	13.7867	14.9535	16.7908	18.4927	20.5992	40.2560	43.7730	46.9792	50.8922	53.6720	59.7031

[a] v = Degrees of freedom.

Index

Milton Keynes UK
Ingram Content Group UK Ltd.
UKHW040447071024
449327UK00020B/1047